Human and Machine Perception 2

Emergence, Attention, and Creativity

Human and Machine Perception 2

Emergence, Attention, and Creativity

Edited by

Virginio Cantoni
University of Pavia
Pavia, Italy

Vito Di Gesù
University of Palermo
Palermo, Italy

Alessandra Setti
University of Pavia
Pavia, Italy

and

Domenico Tegolo
University of Palermo
Palermo, Italy

Springer Science+Business Media, LLC

Proceedings of the Third International Workshop on Human and Machine Perception: Emergence, Attention, and Creativity, held September 14–17, 1998, in Pavia, Italy

ISBN 978-1-4613-7179-3 ISBN 978-1-4615-4809-6 (eBook)
DOI 10.1007/978-1-4615-4809-6

Originally published by Kluwer Academic / Plenum Publishers, New York in 1999
Softcover reprint of the hardcover 1st edition 1999

10 9 8 7 6 5 4 3 2 1

A C.I.P. record for this book is available from the Library of Congress

To Lella and Massimo
thinking very different
as a single person

PREFACE

The following are the proceedings of the Third International Workshop on Human and Machine Perception held in Pavia, Italy, on September 14 - 17, 1998. This edition has been under the auspices of two Institutions: the Cybernetic and Biophysics Group (GNCB) of the Italian National Research Council (CNR), the Centro Interdipartimentale di Scienze Cognitive, of Pavia University and the Centro Interdipartimentale Tecnologie della Conoscenza, of Palermo University. A wide spectrum of topics is covered in this series, ranging from computer perception to psychology and physiology of perception, for analysing and comparing biological and artificial approaches.

The theme of this workshop was focused on "Emergence, Attention and Creativity". The workshop structure consisted of five modules each one composed of two invited lectures (dealing with solutions in nature and machines respectively) and a panel discussion.

The lectures focused on presenting the state-of-the-art and outlining open questions. In particular, they sought to stress links, suggesting possible synergies between different cultural areas. The panel discussion was conceived as a forum for an open debate, briefly introduced by each panellist, and mainly aimed at deeper investigation for the different approaches and strictly related topics. The panellists were asked to prepare a few statements on hot-points, which were delivered in advance to the participants as a guide for a qualified discussion.

The number of participants to the workshop was limited to 50. Besides the 30 invited lecturers and panellists, 20 more participants were admitted. Priority for these positions was given to young researchers who made interesting contributions to the open discussions. Both the lectures and the contents of the panels are included in these proceedings.

Table 1 shows the scheduled program for the five topics. The chapters of the book vary somewhat from the program, to accommodate the positions that emerged from the debates. In what follows the main topics are introduced and each subject is briefly presented and commented.

Table 1. Structure of the workshop in five subjects.

Subject	Nature	Artificial	Panel
Evolution	Biological evolution	Co-operative evolution	Learning and evolution
Emergence	Neuronal focussing and self-organisation	Variability, ambiguity and perceptive systems	Multi-stimuli and action
Attention	Diffused and focussed attention	Alerting mechanism	Hierarchical planning and problem solving
Creativity	Chaos and creativity	Analogy abduction and belief	Creativity in 22 questions
Knowledge	Remembering and forgetting	Representation paradigms	Knowing and reasoning

Topics and contributions

In the last decade, new approaches for machine creativity have been suggested. One of these is based on the Puzzle Principle introduced by Marvin Minsky ("The Society of Mind"); following this approach, any problem can be solved by trial and error provided one already has some way to recognise a solution when one is found. In this connection, new solutions can be detected for problems that we are not able to deal with and to solve with traditional techniques. It is worth to point out two key problems that arise: the generation of candidate solutions, and the evaluation of their proficiency. The highest efficiency is achieved with the capability of perceiving the early emergence of useful solutions. In the workshop we investigated the mutual basic principles subtending Evolution, Emergence, Attention, Creativity and Knowledge. Modalities and capabilities reached or adopted in important natural and artificial contexts were discussed according to the following framework.

Evolution: the outing in constrained open-space. In many applications novelties can be conceived only through a dramatic change with respect to tradition and can spring out by opening the search space and by constraints abatement. Moreover, unexpected outings introduce new approaches and solutions with high gain range; nevertheless, a chaotic search can infrequently be effective, especially when solutions are characterised by very complex patterns. For these reasons the evolution process, from the generation of hints to the final setting of the solution, is a critical one.

Effective evolution requires creativity, and then open-space and high mutation rate in the early phases; meanwhile requires precision, then introducing constraints and low mutation rate or even just selection in order to optimise, in the final phases. The agents of the lower levels act as artists or inventors, producing many ideas ('sturm und drang' phase); the most promising ideas are then passed to the drafters or engineers at the higher levels were they are tested and if successful made more precise.

Particular interest is given to the evolution of self-organising or collective and co-operative systems and populations. In these cases, two hierarchical levels of interaction and adaptation can be considered: the evolution and adaptation of populations with their environment, and at a lower hierarchical level the adaptation of individual within those populations. In this connection evolution can be seen as the capacity for change with improvement in fit by learning through the interaction with the environment.

Emergence: the issue of change. In its emerging meaning evolution is related to things that change by emerging, persisting, or decaying in their environment. The absence of dynamic has no by-product; instead change creates opportunity and devastation alike. The effective exploitation of opportunities emerging from a complex and a chaotic environment depends on how early they are detected and on responding swiftly and in appropriate way.

Attitude to innovation demands the capability to detect structure and order in the chaos. Two proficiencies characterise this attitude: the willingness to look on change as an opportunity and the art of catching positive seeds. Selection accounts for the emergence of useful specificity, as for biological evolution that produces classes of living from the bottom-up by gradual selective processes (over eons of time). Diversity exist beforehand in a selective system, specificity exists as a result of the selection 'ex post facto'.

There is not explicit information transfer between the environment and the individual elements that cause the population to change and increase its fitness. Anyway, even in autonomously evolving system, the interaction between the observer and the system must be taken into account; this either in case of endogenous competing evolutionary processes or in case of evoked changes (e.g. heuristic production of candidate solutions).

Attention: the trend-forecasting tool. Attention is a generic concept that has to be defined depending on the system knowledge concerning the task at hand. Moreover, attentive mechanisms must be applied whenever the amount of data is much larger with respect to the information necessary to solve a problem. For each candidate solution an analysis must be done to determine pros and cons, and the reliability of the foresight. In most practical cases the solution space is huge and requires for massive computation. A supervisor for exploratory mechanisms is essential to tackle the complexity of the analysis. Let us quote as a critical example the Internet chaos and the *data mining* techniques.

Selective or diffused attentive strategies can be adopted to solve the cases in which the expected events are known or in surveillance over the whole parameter space, respectively. Both cases have been considered in the workshop, in particular with regards to the lower level for the role of alerting mechanism in the visual system (early vision stage and the saccadic exploration). At the higher level, instead, regards to the internal models and/or semantic knowledge that advice and address our sensors to select *useful* information.

Practical applications are also considered, namely in robotic navigation and car plate recognition, socio-technical systems, co-operative players, and planning the attention in problem solving.

Creativity: the power of positive dreaming. The production of novel and original issues depends on the capability of observing the context (seeing through to the essence of things, looking at one thing and seeing something else in it, inspiring 'lateral' thinking, etc.) but benefits as well from the readiness to discover. Even in this case, two proficiencies characterise this attitude: imagination up to dream and a prepared mind (from Pasteur: "*chance favours the prepared mind*").

In general, a free evolution would provide from the maximum novelty expectations: the stronger the interactions, the lower the originality of the outcomes. Nevertheless, constraints can hinder but also foster creativity: in many cases overt or explicit constraints can act as creativity catalysers by focussing the attention on the critical essence.

According to Johnson Laird, an operational definition of creativity can be derived from three properties: i) it is generated from previous knowledge; ii) it isn't restricted to reach a precise goal (freedom of choice); iii) a creative process is the result of a non-deterministic procedure (unexpected result).

Creativity can be the result of evolution processes and many times is based on the perception of the analogy with the approach to the solution of a different problem. Associative links favour the discovery of hidden gemstones 'out there, waiting to be found'. The increasing availability of structured information allows better associative correlation in large amount of data, allowing induction and abduction processes as the basis for new model of reasoning.

Knowledge: the ascertain of salient and scanty information. The (biological or artificial) system knowledge is related to the capability of interaction of the system with the environment. The genetic information consents to humans to start this interaction; knowledge is then enriched by facts and notions, as well as by their relationships as they are acquired from daily experience. These evolution and maturation processes consist on passing from genotype into phenotype.

Two key points for knowledge usage discussed in the workshop are coding and retrieval processes. Another issue regards the notion of consciousness as a basic aspect of information knowledge that makes difficult the implementation of knowledge management in artificial systems. Nevertheless, it musts be pointed out that it is also questionable the presence of consciousness in animals. And even for humans, according to the Gerbino panel report *"The problem regards the means for representing to ourselves the content of our own thoughts.... "*.

This point is well understood in teaching class where the transfer of knowledge to students can be strongly increased by the teaching tools (from oral communication, to power point presentation, to movies on simulations of phenomena, etc.). Visual information permeates our activities; communication and representation is becoming more and more relevant.

The market in computer engineering is driving the researches towards intensive analysis of the quality of computer-user communication, and the centres which measure quantitatively the effectiveness and the usability of human machine interaction are rising everywhere. Moreover, machine-machine communication is becoming more and more effective, including also sophisticated techniques of concurrency and co-operation between artificial systems.

Acknowledgements

The workshop, and thus indirectly this book, was made possible through the generous financial support of the universities involved, and the research and industrial organisations that are listed separately. Their support is gratefully acknowledged.

The editors would also like to express their appreciation to the Program Committee of the workshop for their advice: Ornella Andreani-Dentici, Jörg D. Becker, Roberto Bordogna, Luciano da Fontoura Costa, Egidio D'Angelo, Ivo De Lotto, Salvatore Gaglio, Peter Gärdenfors, Giorgio Grisanti, Eugenio Mira, Silvana Vallerga, Yehezkel Yeshurun, Bertrand Zavidovique.

Special thanks should go to Paola Bernardi and the people of the Collegio Nuovo, for their precious help in organising the workshop.

Virginio Cantoni, Vito Di Gesù, Alessandra Setti, and Domenico Tegolo

SPONSORING INSTITUTIONS

The following institutions are gratefully acknowledged for their contributions and support to the Workshop:

- Gruppo Nazionale di Cibernetica e Biofisica del C.N.R.
- CARIPLO
- Università degli Studi di Pavia
- Comitato 7 (Scienze d'Ingegneria e Architettura) del C.N.R.
- Elsag Bailey
- Dipartimento di Informatica e Sistemistica dell'Università degli Studi di Pavia
- Banca Regionale Europea
- Azienda Promozione Turistica di Pavia
- Centro Interdipartimentale di Scienze Cognitive of Pavia University
- Centro Interdipartimentale di Tecnologie della Conoscenza of Palermo University

CONTENTS

EVOLUTION

EMERGENCE

ATTENTION

CREATIVITY

KNOWLEDGE MANAGEMENT AND REPRESENTATION

THE DYNAMICS OF BIOLOGICAL EVOLUTION AND THE IMPORTANCE OF SPATIAL RELATIONS AND SHAPES

Luciano da Fontoura Costa and Luís Augusto Consularo

Cybernetic Vision Research Group, IFSC-USP
Caixa Postal 369, São Carlos, SP, Brazil
FAX: +55 16 271 3616
e-mail: luciano@if.sc.usp.br
http://www.ifsc.sc.usp.br/visao/group/members/luciano/luciano.htm

Creativity consists in seeing what everyone else has seen and thinking what no one else has thought.
(Adapted from a quotation by A. Szent-Gyorgi)

INTRODUCTION — *Homo Scientificus*

Since the beginning of times, matter and energy have been flowing throughout the universe. Yet, instead of following a dull progression (such as dust in the wind), our universe has exhibited an inherent tendency toward a most exuberant dynamic evolution. Starting with elementary particles such as quarks, matter and energy have inexorably aged towards increasing complexity, giving rise to life and even humans, the most exotic and complex known outcome of such a process. Part of this chapter addresses the important and intertwined issues of *life* and *evolution.*

Human beings are exotic to such an extent that we started asking fundamental questions about ourselves, the universe, and our position with respect to the latter (...and psychoanalysis was born). As a matter of fact, the tendency of making difficult fundamental questions — who are we? what is our role in the universe? what is the universe made of? what is life and death? — are perhaps the most quintessential hallmarks of being human. Out of the many attempts tried by humans to answer such questions, including the use of oracles, science has been the most successful one. Though limited and imperfect, it is still our very best way to treat the fundamental questions which have haunted humankind, being actually its direct byproduct. While science does not guarantee solution — nature does not come with guarantees[1] — or offer a smooth path to discovery, its success has lied in its inbuilt error detection and correction mechanisms.[2] Scientific advances often begin with hypotheses and insights. While hypotheses provide some kind of levers (in Aristotle's sense), and are by definition something that can not be proved, insights exhibit a more subjective and even "magical" nature. Yet, insight is the most

Human and Machine Perception 2: Emergence, Attention, and Creativity,
edited by Cantoni *et al.* Kluwer Academic / Plenum Publishers, New York, 1999.

dynamic and decisive element in science, being directly related to *creativity*. Since science generally does not point out the smooth path to solution, but mostly provides a means of verifying the validity of initial guesses, it is of paramount importance to start with good insights. Having such an importance, it should not be surprising that insight has become itself a subject of scientific inquiry. The great prize at stake here is the capability of understanding how insight is generated, and using this knowledge to speed up the process of scientific discovery. This is a quest for catalysts of science. Part of the present chapter is about *creativity*.

Science operates by conceiving models which can explain and predict nature and by assessing the achieved successes. In a sense, we are building oracles, but of a more rational nature, embedding error detection and correction mechanisms. Oracles which are expected not only to help us to understand the universe, but also predict its future. The power of science sometimes places humans close to god, but without any guarantee of complete success.

Being a product of the evolutionary process, human perception has been optimised and suited to the environment and problems typically met by humans and, as a consequence, provides only a rather constrained and biased window to the universe. For instance, we are inherently visual creatures, even if restricted to the visible range of the electromagnetic spectrum. In addition, our most immediate perception of the world is tuned to the human spatial scale, ranging roughly from millimetres to meters. No wonder we have always had difficulties in grasping extreme values (e.g. astronomical and microscopic). More recent evolutionary features, such as social behaviour, language and culture, have contributed further to strongly biasing and constraining our perception of the universe and shaping our intuitions to such an extent that it is often difficult for someone from a specific culture to understand and interact with stranger cultures. As Plato already knew, our perception provides only a distorted glance at the *world of ideas*, which will be henceforth understood as mathematics and geometry. Yet, throughout a painstaking process lasting thousands of years, humans have succeeded reasonably well in overcoming such biased perception and explaining, in terms of models based on the *world of the ideas*, a good number of phenomena in the known universe. Interestingly, the first breakthroughs were achieved where human intuition was close to nature, the subsequent developments becoming more and more difficult since they departed considerably from intuition. Take classical and quantum mechanics, for instance. While the development of the former already required a substantial degree of insight, perhaps with some help of apples, the latter presents a predominantly non-intuitive and abstract nature, being only too difficult to be conceptually mastered by humans. We believe quantum mechanics because it is a physical model which works well, allowing good explanations and predictions. In fact, the informatic revolution would not have been possible without quantum mechanics. However, while many of the fundamental questions remain unanswered, we have no doubt made immense advances at least in posing new questions, since each new answer brings with it a series of new questions. There can be little doubt that, if being human corresponds to asking fundamental questions, we have been performing with flying colours.

But there is another important aspect to science: that of models which can never be complete. For instance, when a physicist models a pendulum, he has to identify and concentrate on the most important natural features, ignoring an infinity of secondary effects (e.g. the gravitational field of not only distant galaxies, but even of our closest neighbour, the moon). Models in science are thus cursed to be incomplete. Since non-linear effects can amplify small perturbations, and even lead to chaos, the selection of the elements to be included in a model is a most critical endeavour. Scientists are not only expected to select suitable hypotheses and be creative, but also to identify the suitable elements for their models and, of course, these two endeavours are inter-related. To be more specific, the identification of the relevant elements to be included in a model is a function of the degree of desired explaining and predicting powers, in such a way that simpler models tend to present more limited capabilities. As a matter of fact, such selective process is also inherent in choosing the hypotheses. Be that as it may, selecting elements for scientific models involves an important and very human process, namely *attention.* In other words, given the infinite richness and complexity of nature, scientific models can be understood as means of paying attention to specific elements of the universe, hopefully those which will lead us to effective explanations and predictions. To discuss *attention* provides another subject for the present chapter.

As it becomes clear from the above lines, the scientific process, born out of our selective attempts to explain and predict the universe, presents a series of parallels with the highest human features, since every human is to some extent a scientist. Creativity becomes a most fundamental element for generating new perspectives and solving problems, and attention arises from the fact that we can never be completely comprehensive about the universe. The first part of the present chapter (second section) corresponds to an attempt at revising and discussing in an integrative fashion the concepts of life, evolution, creativity, and attention. Since such issues have been extensively treated elsewhere, our efforts are focused on the following two main attempts: (i) to review, from a point of view as much as possible detached from the human biased perspectives, the concepts of life, evolution, creativity and attention; and (ii) consider spatial relations and shapes as important overlooked parameters heavily constraining and influencing these four concepts. This chapter also provides two case-examples illustrating such possibilities with respect to the relationship between neural shapes and behaviour, and visual attention.

LIFE AND EVOLUTION

The definition of life has been highly controversial.[3] While many of the attempted approaches often disagree at some point,[4] they typically involve concepts such as:

(a) living beings must die and/or decay (Aristotle);
(b) life is restricted (or not) to chemical substrates;
(c) living creatures are capable of self-reproduction;
(d) living beings evolve (e.g. towards increasing complexity);
(e) life involves metabolic and homeostatic processes;
(f) life depends on *"élan vital"*.

While not intended to give the final answer to what life is, we shall try in this section to develop a definition which may serve as a working hypotheses for the rest of the present chapter. After all, at least from human perspective, evolution, creativity and attention are all properties closely (but perhaps not exclusively) related to life. Let us start by the close relationship between life and evolution.

Since it seems certain to us that living beings in our planet present an inexorable tendency to evolution, life itself does not seem to be dependent on everlasting evolution. For instance, we have the case of bacteria and viruses, which have been exceedingly successful and stable forms of life for millions of years. Indeed, this observation brings into focus another important element: in what sense can we say living creatures are *successful*? What is the *fitness function* for life?

Before addressing such a fundamental question, let us start by the general assumption that life is a property exhibited by *individuals while interacting with their respective environments*. Though we do not know of any living being not fitting such an assumption, defining individuals can be a subtle issue in some cases, such as with algii and coelenterates, especially regarding their spatial boundaries. One of the most effective means of spatially constraining individuals is in terms of *cells*, one of the basic, but not exclusive, building blocks of more complex living creatures. Next to self-replicating molecules and viruses, procaryotic cells are the most primitive known forms of life. As it is currently believed, these first cells originated when primitive self-replicating molecules become involved by bilayer membranes, which can be formed in water with relative facility as consequence of electric interaction between polar water and amphipatic molecules such as phospholipids.[5] The quintessential role of membranes is to provide a good degree of isolation between two media, such as the interior of the cells and the rest of the environment. Such isolated environments are believed[5] to have provided a more controlled and stable internal spaces where self-replication could suffer less interventions and where useful specific molecules, such as proteins, could be kept and help the process.

Membranes provides a reasonably clear-cut boundaries to cellular individuals. But only reasonably, because proteins embedded in the membrane provide extensions of the direct interactions between cell and environment, and fields extending from and into the cell are also strongly connected to individuals. It is interesting to point out that the next evolutionary step involved the addition of a further isolating membrane inside the cell, in order to separate the genetic material (in the nucleus) from the rest of a more complex cell machinery, originating the eucaryotic cells. Now, while procaryotic cells have evolved into intricate solutions to life, involving cilia, flagella and the most complex morphologies, and even primitive attempts at social organisation (e.g. myxobacteria[6]), eucaryotic cells present a more definite tendency to interact and form agglomerates. Two of the most basic possible advantages of such a strategy are: (a) the capability of becoming larger allows better sampling the environment for food and mates, and also for sensing dangerous elements; and (b) the possibility of evolving more complex behaviour by sharing tasks between specialised (or not) cells. An interesting example of primitive eucaryotic colonies is the *Gonium*, corresponding to a number (actually a positive integer power of two) of cells adhered to a matrix of extracellular molecules.[5] Since the cells in this creature present flagella on a same side, the whole colony becomes capable of certain motility, allowing a more effective sampling of the environment.

It should be observed that the emergence of such composite creatures present some problems in spatially constraining individuals. For instance, there are cases of living masses involving more than a single distinct species (e.g. symbiosis), and there are also agglomerates of cells which, though including only a same species, do not exhibit strong interactions or definite morphology allowing a clear definition of the individual. Consequently, we shall henceforth constrain ourselves to calling individuals those masses of cells, all of a single species, exhibiting a certain minimum degree of energy and mass exchanges between themselves and with the environment.

Having worked out a reasonable definition of individual, an important next step is to investigate and characterise its interaction with the surrounding environment. In order to perform its basic needs, uni- and multicellular individuals have to constantly exchange mass and energy with the environment (open systems).[7] Such interactions are selective, in the sense that only some of the elements around are vital for each creature, entailing the need to seek and recognise the needed material. It is interesting to note that such a filtering is common both regarding energy (e.g. filtering specific wavelengths of electromagnetic radiation) and mass (e.g. searching for specific proteins) interactions, which are highly dependent on the specific *shape* of creatures. Ample surfaces, such as those presented by leaves, become essential for effectively capturing solar energy. Large surfaces are also a bonus for creatures sampling the environment for food, such as sponges, medusa, and even whales (their widely open mouths). But spatial coverage, which can be also accomplished through intricated ramified structures, is not the only important aspect related to interactions between individuals and the environment, for many elements in the 3-D environment space propagate their presence through scalar and vector fields extending over very long extensions. Though all matter affects space through the gravitational field, this influence is of very limited ecological value since most creatures lack specific sensors.

More relevant interactions include electric and optic fields, as well as diffusions of chemical elements, ions, molecules, and even food (e.g. shoal of fish). Yet another important role performed by the shape of living creatures concerns the displacement of these individuals, in aero and hydrodynamic terms. Perhaps the role of spatial relations and shapes, so essential in defining field interaction between individuals and the environment, have been relatively overlooked in studies of life and evolution.

ATTENTION

Since models are necessarily limited and provide but simplified representations of the universe, they can be understood as a means of paying attention to specific features of the environment. Even individuals could be thought as means of paying attention through constrained interactions with their environment.

The key concept underlying attention is *selection.* Human beings are to a great extent attentive creatures, and even our conscious awareness presents (or at least seems to) a sequential flow, in such a way that we usually process only a thing at a time. As observed above, attention seems to entail a limitation of the interaction between some object and its environment, as a means of allowing or making some process more effective. In the case of cells, only specific interactions — namely those which have survival value — are allowed as a means of ensuring better control over the processes taking place within the membrane-bound space.

It is argued in this chapter that the concept of attention at large can be generalised from the above situation, i.e. attention is the process of constraining (or filtering) interactions in order to favour some specific process. Attention is important not only for avoiding harmful exchanges, but also as a means of optimising energy and physical resources. Take visual attention, for instance. Were humans to process the whole visual field with the same spatial accuracy, without concentrating attention in the foveal region, we would have to have enormous cortical regions which would not fit into the head.

A big question entailed by attentive mechanisms concerns how to select the aspects that are more important, in space amongst many components, and sequentially in time. It becomes clear from the above definition that such a selection must be determined in order to favor those interactions leading to more effective performance of the process in question. A possible means of doing so is through the evolutionary mechanism.

CREATIVITY

Creativity can be thought as the means of obtaining good solutions, which are particularly effective, and often surprising or uncommon, to some specific problem. Though being a relatively simple definition, the terms "good", "particularly effective, surprising and uncommon solutions" demand some additional considerations. Let start with "good". This requirement entails some fitness metric, in the sense that a solution can be considered good only with respect to some specific relative objectives. For instance, turning on a light in a living room can be a good solution while searching for something, but is definitively a bad solution for cooling the environment. On the other hand, the problem of achieving effective selective interactions between individuals and their environment could also be thought as a sort of natural creativity.

“Particularly effective” and “surprising/uncommon solutions” are meant to exclude those solutions which are awkward, trivial and/or effort-consuming. Such requirements no doubt account for the *difficult* element behind creativity. To be more specific, which sort of procedures can be devised for pointing out effective solutions to specific problems?

Formally speaking, these issues bear close relations to the computability of specific problems. NP-complete problems, for instance, are not very likely to be solvable by effective algorithms. How can one be creative in such cases? Being this a most difficult question, let us concentrate on more amenable cases involving creativity, especially those related to human activities and science.

Let us return to the "surprising/uncommon" element in our working definition of creativity. It is a matter of fact that inventive processes, including the great breakthroughs in science, are often accompanied by completely different perspectives and non-intuitive approaches to traditional problems. Quantum mechanics provides an excellent example, in the sense of modelling particles as clouds of probabilities rather than exact speed and position, a completely non-intuitive concept. It soon transpires that being creative involves detaching oneself from the human perspective and intuitions which have been impinged onto us by our specific evolutionary progress. To see things in a completely different and innovative way — that is for certain an important element in creativity and insight.[8] Yet, how can we get completely rid of our biased perception? Moreover, would that be wise? Perhaps the more sensible alternative, at least at a more short term perspective, is to best explore and apply our limited and biased perception and intuitions, trying to develop and expand it in such a process. It should be observed that by allowing creativity to involve either an uncommon or surprising nature, it is possible to associate creativity to both the optimisation processes in nature (uncommon solutions), and human insights (surprising solutions).

As with attention, shape and spatial relations are also important for creativity. For instance, it is possible to combine a triangle into many more distinct objects that would be possible with a circle.[8] In other words, higher shape complexity leads to a large space for creativity. Also, at least in humans, visual creativity in combining shapes is largely performed by paying attention to salient features,[9] such as the vertices of a triangle.

TOWARDS AN INTEGRATED APPROACH

In the second section a preliminary definition of *life* was advanced as the property allowing individuals, which are themselves filtered portions of the environment resulting from a long evolutionary process, to interact in an adaptive way with ever changing surrounding environments through specific energy and mass exchanges constrained by physical factors, among which spatial relationships and shapes seem to play a particularly important, if relatively overlooked, role. Such adaptation capabilities can be accomplished by incorporating mechanisms that allow the individual to change itself, and/or by generating new versions of suitably modified descendents which can better cope with new environments (we shall refer to these two processes as *intelligent* and *brute force*, respectively). In other words, enduring life in nature is only possible through *evolution*, which we shall understand as any process allowing living individuals to adapt to their changing environments, through selective interactions, in order to guarantee a extended span of life.

In life as we know in this planet, while inbuilt adaptive capabilities allow a limited degree of adaptation to environmental changes, coping with large changes can only be dealt with through the brute force method, i.e. by being able to produce varied offspring. Yet, it is important to note that self-modification by itself could, at least in principle, be enough, though it would involve a highly complex inbuilt adapting machinery capable of coping with drastic changes. This observation has profound implications. In order to effectively coping with changing environmental conditions, which *include other competing individuals*, each living being has to rely on *models* of the environment allowing them to predict the future to a certain extent in order to better interact with the environment.

However, more and more complete models are only possible in more and more complex individuals, hence the evolutionary tendency towards increased complexity. On the other hand, increased complexity entails higher metabolism (our brains consume a great deal of our energy), thus acting as a brake in the process. Be that as it may, we are a most remarkable example of such a tendency, for our evolution has consisted of the gradual but inexorable replacement of brute force by more intelligent and flexible self-modifying capabilities allowed by our ability to develop and handle more complete models of the world. In our present form, most of our evolution is accomplished through the latter strategy, and human evolution has become governed by culture and science, and not intense reproduction aimed at varied offspring. We have not (yet) come all the way from chemistry into thought based evolution, but became capable of interceding in the very chemical basis of our lives and, perhaps even capable of creating and designing life in the future.

It is now important to return to the question regarding how we can rate the success of life. It is now clear that adaptability seems to be the key issue, for it allows longer spans of life and more complete interactions with the surrounding environment, especially regarding competition between individuals. Perhaps a reasonable fitness function for life should be one that expresses such a degree of adaptability, which is itself closely linked with the individual's ability of modelling the world in a selective and profitable way. As Carl Sagan so nicely expressed, humans may indeed be means of the Cosmos becoming conscious about itself - though, as we would like to add, still in the imperfect, incomplete and biased manner allowed by our limited models of the universe.

There are some intriguing direct analogies between natural evolution and the scientific character of being human in the sense that both endeavours involve deriving models of the world to be used for ensuring life continuation in a changing universe in the former case, and to answering what life and the universe is in the latter. That these two quests are closely related may provide the seed for explaining both.

Considering the concepts and definitions advanced thus far in this chapter, it is tempting to propose an integrated view of life, individuals, attention and creativity. In short, evolution can be thought as the process of ensuring good adaptability of individuals in their environments, specially by developing models of the world which are more and more complete. Being necessarily constrained, models involve paying attention to specific features of the universe, a process which has to be effective in order not to jeopardise individuals while competing with others during the evolutionary process. The achievement of such effective models and selected interactions with the environment can be obtained through "creative" processes acting in the natural world.

Having thus worked out an integrated approach involving the concepts of life, evolution, attention and creativity, it is time to proceed to the second major issue addressed in the present chapter, namely the importance of spatial relations and shapes to the above mentioned concepts. The basic idea here is to consider influence fields, scalar or vector, as particularly suitable means of characterising, analysing and modelling processes such as attention and creativity. One of the important properties allowed by such an approach is that influence fields allows us to clearly express long and short term spatial interactions. Moreover, the distribution of the lines of force (converging to the surface of objects) in such fields provide a nice means for expressing how the surrounding space is covered by the objects. The following two sections provide two case-examples exploring such possibilities.

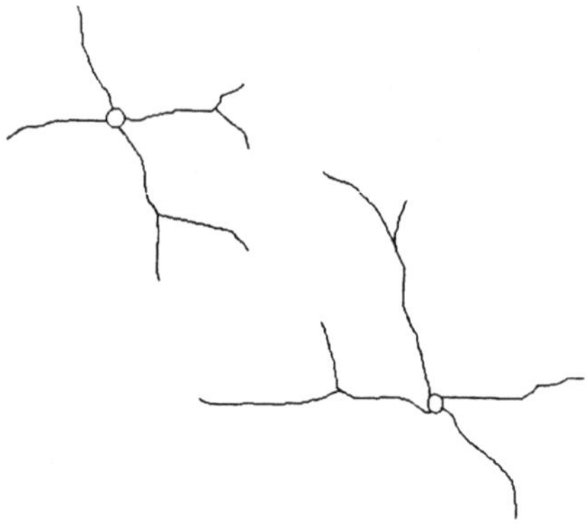

Figure 1. The two neural cells considered herein.

CASE_EXAMPLE 1: RELATING NEURAL SHAPE AND BEHAVIOUR

In nature, and particularly primates, neurones come in the most diverse shapes and sizes. Considering the limited and already saturated capability of genes in defining the phenotype, it would be most surprising if the shapes of neurones, which are at least partially defined by the genes, were purely accidental and not related to neural behaviour. Indeed, the outstanding and pioneering work of Santiago Ramon y Cajal, along the end of the last century and beginning of the current one, concentrated in identifying the intriguing variety of shapes of neurones. That researcher, often considered the founder of modern neuroscience and up to the present day the most cited neuroscientist, went so far as to explain the whole of human complex behaviour to the shapes of neurons.[10] Interestingly, and partially as a consequence of the advances in electrophysiology, the developments in neuroscience along the current century have mostly concentrated in electrophysiological experiments and modelling approaches, such as the transmission cable and Hodgkin and Huxley models.[11]

It is the belief of the first author of the present chapter that neural shape and behaviour are closely intertwined, strongly affecting each other.[10,13,14] This interaction can be perhaps

Figure 2. The scalar field defined by the cells in Figure 1.

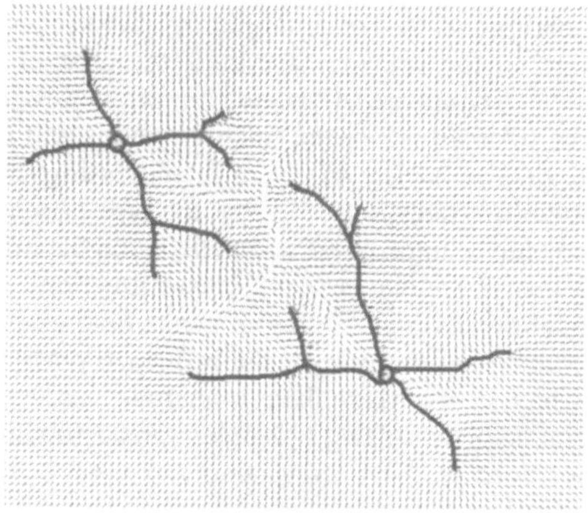

Figure 3. The vector field respective to the cells in Figure 1.

best understood in terms of the shape imposing constraints to the biochemical processes going on inside the neural cells. A spherical cell, for instance, would exhibit a much more simple behaviour than another with more evolved processes. Of course, the additional complexity would not be directly related to the synaptic effectiveness, but the electrotonic and active processes going on along the dendrites and axons. But shape and relative position may actually be decisive also at the synaptic level, for the synapses are established in terms of influence fields and the relative spatial position and shape of the cells. The growth cones[15] which pioneer the synaptic pathways are known to be guided by a combination of long and short range interactions, such as electric field and ionic concentration gradients. As a matter of fact, the topographic maps present in most of the primary cortical areas of primates,[16-18] itself a feature which can only be achieved through a painstaking and careful dynamic process occurring during the early stages of development of the animal, provide additional evidence that the relative position of the cells is important for behaviour. This section describes some computational approaches

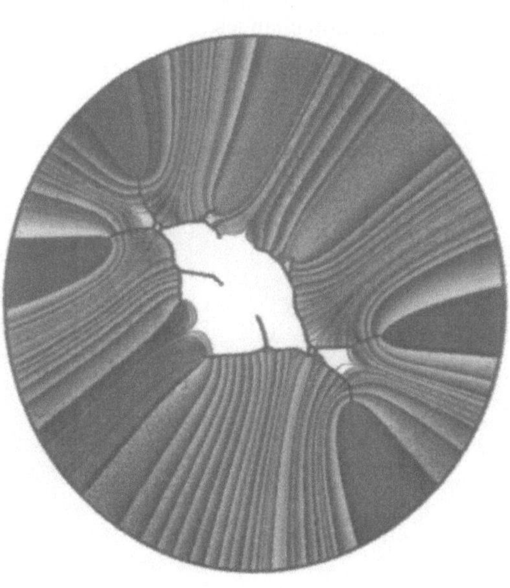

Figure 4. Lines of force around the cells in Figure 1.

and results illustrating how neural shape and function can be related.

In a way that is analogous to the interactions between individuals and their environment, as discussed in the previous sections, the area of influence and spatial coverage of neural cells[18-20] should be of special importance regarding the interaction between neurones and the intercellular space (including ions and neurotransmitters) and even other neurones and glial cells. The first step in addressing this problem consists in defining the *type* of influence in question.[12,22] Once the type of interaction is defined, we can think of a neurone as being a single point, and then obtain the influence profile that cell would induce (in optics this is known as point spread function). Now, through a well-known theorem from signal processing,[23,22] the influence field established by the neurone can be expressed by the convolution between the influence profile and the shape of the neurone (i.e. a function which is one where there is neurone and zero otherwise), a process which, given the extensive support of the 2-D point spread function, can be effectively performed in the Fourier domain.[24] By using a similar approach, it is also possible to determine vector fields of influence (e.g. electrical field) and even analyse and measure the density of lines of force reaching the neurone membrane.[22,13] The remaining of the present section illustrates such possibilities with respect to the 2-D neural cells depicted in Figure 1. It should be observed that many neurones in nature exhibit a predominantly planar organisation, being suitably represented as 2-D shapes.

Figure 2 presents the scalar field generated by such cells assuming inverse square intensity decay (such as in potential/electric fields). Each point of the 2-D shape is treated as a unitary charge extending to infinity in both directions along the z-direction (i.e. orthogonally to the plane of the cell). The potential of such graphical representations to express the influence produced by the cell over the surrounding space is self-evident. Figure 3 exhibits the respective vector field, and Figure 4 illustrates the lines of force (continuous curves which are always tangent to the vector field) originating in a circle centred at the cells. Such representations may provide important subsidy for understanding how neural interconnections evolve, and the density of lines of force reaching each point in the cell membrane can give us an idea of the change of having synapses at each point (concentration of converging paths). Such a density, which in conservative fields only makes sense because the space is quantised (otherwise the lines would never meet), is expressed in terms of histograms in Figure 5. In addition, such field interactions may provide the key for expressing and modelling dynamic interactions between neural cells based on electric and ionic concentration changes.

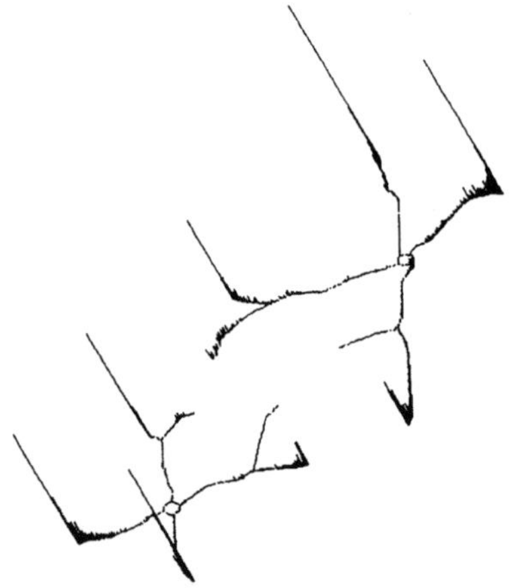

Figure 5. Histograms of the number of lines of force reaching each point along the spatially quantised cell contour.

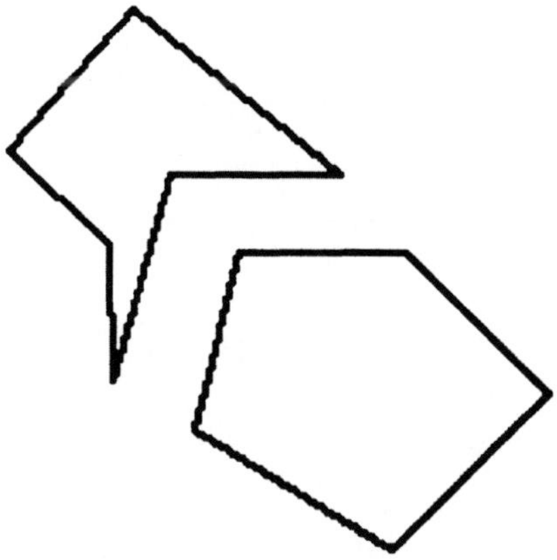

Figure 6. Two geometrical contours.

It should be observed that all these approaches can be directly extended to 3-D cells. Additional evidences regarding the interplay between neural shape and behaviour, obtained through simulations of orientation encoding in centric neural domains with cell bodies distributed according to a grid and randomly, have been identified by the first author regarding gradients of the size of dendritic arborizations.[22,24]

CASE-EXAMPLE 2: SPATIAL RELATIONSHIPS AND SHAPES AND SELECTIVE VISUAL ATTENTION

The concept of fields of influence for analysing and modelling interactions between objects in space can be directly extended to visual perception, as discussed in this section.

It has been known for a long time[9] that corners and high curvature points along 2-D shapes act as special attractors of visual attention. In other words, given a triangle, our

Figure 7. Scalar field for the objects in Figure 6.

Figure 8. Lines of force for the objects in Figure 6.

attention is bound to concentrate on its vertices. The concepts of scalar and vector fields, as well as lines of force and their density along the contour, can all be directly applied in an attempt at modelling the important phenomenon of visual attention. Figure 6 presents two geometrical shapes hereby used to illustrate such an approach. The respective scalar field is shown in Figure 7. Figure 8 presents the lines of force, and Figure 9 shows the histogram of lines of force. It is also possible, and we have been investigating it, that the lines of force may be correlated to the pathways performed by our visual analysis of the shapes. Though such a formulation could account only partially for the complex problem of visual attention, the first author has identified, as part of the M.Sc. work by Portezani,[26] a relatively good agreement between the points selected by human subjects and the point to which the lines of force originating at the initial fixation point converges.

CONCLUDING REMARKS

This chapter was aimed at providing an integrated discussion about life, individuals, environment, creativity and attention, while paying special attention to the important issue of spatial relations and shapes. Evolution has been presented as a consequence of the need to cope with changing environments, and specially the continuing competition between different individuals, which imply a more dynamic process in which models of the environment have to be constructed. Being necessarily incomplete, such models act as a sort of attention, in the sense that specific relevant features of the environment have to be identified and selected. Considering such a scenario, creativity in nature can be thought as a means of achieving effective attentive processes favouring interactions between individuals and their environments. On the other hand, human creativity can be thought as the process of finding particularly effective and surprising solutions despite the bias imposed by the intuitions of the individuals. Being a sort of optimisation process depending on a specific fitness function, creativity can be achieved in machine, for instance by using the genetic algorithm. Thus, life and evolution have in great part been closely related to creativity and attention.

In the second part of the present chapter, it was argued that the interaction between individuals, including neural cells, can be properly described and modelled in terms of field interactions, which are governed by the spatial relationship between individuals as well as their respective shapes. Some approaches at characterising and modelling such

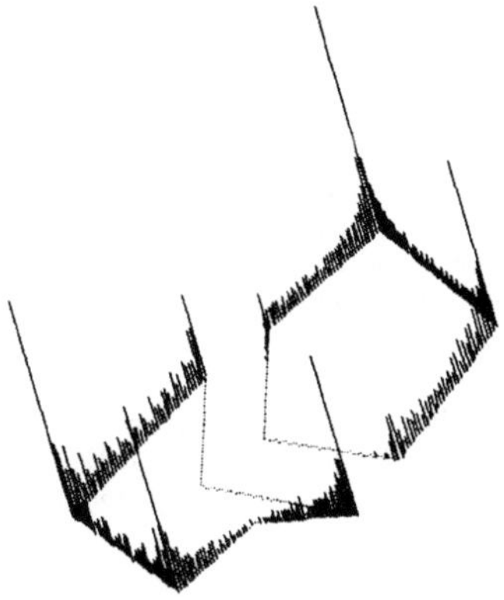

Figure 9. Histograms of lines of force for the objects in Figure 6.

interactions, having in mind the specific cases of neural structures and visual attention, have been presented and illustrated through some examples.

As the near century (and, as a matter of fact, millenium) approaches, it is hoped that we will be returning to the ideas firstly advanced by Cajal at the outset of the present century, and paying more attention to shapes and spatial relationships.

ACKNOWLEDGEMENTS

Luciano da F. Costa is grateful to FAPESP (Procs. # 96/05497-3 and 94/04691-5) and CNPq (Proc. # 301422/92-3) for financial help. Luís Augusto Consularo is grateful to CNPq (Proc. #143725/96-3) for financial support.

REFERENCES

1. B. J. Baars, *In the Theater of Consciousness: The Workspace of the Mind*, Oxford University Press (1997).
2. C. Sagan, *Billions and Billions: Thoughts on Life and Death at the Brink of the Millenium*, Random House (1997).
3. S. Franklin, *Artificial Minds*, The MIT Press (1995).
4. S. Levy, *Artificial Life*, Vintage Books (1993).
5. B. Alberts, D. Bray, J. Lewis, M. Raff, K. Roberts, and J.D. Watson, *Molecular Biology of the Cell*, Garland Publishing (1994).
6. E.P. Solomon, L.R. Berg, D.W. Martin, and C. Vilee, *Biology*, Saunders College Publishing (1993).
7. S. Kauffman, *At Home in the Universe*, Oxford University Press (1995).
8. R.J. Sternberg and J.E.. Davidson, *The Nature of Insight*, The MIT Press (1995).
9. F. Attneave, Some Informational Aspects of Visual Perception, *Psychological Review*, 61:183 (1954).
10. S. Ramon y Cajal, *Recollections of My Life*, MIT Press (1989).
11. J.M. Bower and D. Beeman, *The Book of GENESIS*, Springer-Verlag, New York (1995).
12. L. da F. Costa, New Perspectives in Neuromorphometry and Computational Neuroscience, in: *SIBGRAPI'97 site*, http://www.visgraf.impa.br/sibgrapi97/invited/costa.html (1997).
13. R.M. Cesar Jr. and L. da F. Costa, Semi-Automated Dendrogram Generation for Neural Shape Analysis, in: *Proceedings of SIBGRAPI'97*, L.F.Figueiredo and M.Lobo Netto eds., IEEE Computer Society, Los Alamitos, CA (1997).
14. L. da F. Costa and T. Velte, Automatic characterization and classification of ganglion cells from the salamander retina, *J. Comparative Neurology*, (1998) in press.
15. R.V. Stirling and S.A. Dunlop, The dance of growth-cones - where to next?, *TINS*, 18:11 (1995).
16. D.H. Hubel, *Eye, Brain and Vision*, W. H. Freeman and Co (1995).
17. S. Zeki, *A Vision of the Brain*, Blackwell Science Inc (1993).

18. A. Peters and E.G. Jones, *Cerebral Cortex*, Plenum Press (1985).
19. D.M. Dacey, The mosaic of midget ganglion cells in the human retina, *J. Neuroscience*, 13:5334 (1993).
20. J.J. Stein, S.A. Johnson, and D.M. Bergson, Distribution and coverage of beta cells in the cat retina, *J. Comparative Neurology*, 372:597 (1996).
21. S. Vallerga and C. Usai, Relation between light responses and dendritic branching in the salamander ganglion cells, *Experimental Biology*, 45:81 (1986).
22. L. da F. Costa, R.M. Cesar Jr., R.C. Coelho, and J.S. Tanaka, Perspective on the Analysis and Synthesis of Morphologically Realistic Neural Networks, in: *Modeling in the Neurosciences: From Ionic Channels to Neural Networks*, R. Poznanski ed., Gordon and Breach Science Publishers (1998) in press.
23. K.R. Castleman, *Digital Image Processing*, Prentice Hall, Englewood Cliffs-NJ (1996).
24. L. da F. Costa, Gauss' law in image processing and analysis via fast numerical calculation of vector fields, in: *Journal of Real-Time Imaging*, Academic Press (1998) in press.
25. R. Coelho, and L. da F. Costa, Morphologically realistic neural networks, in: *Proc. Third International Conference on Engineering of Complex Computer Systems*, IEEE Computer Society (1997).
26. C. Portezani, *Modelagem de Atenção Seletiva usando Campos Vetoriais*, M.Sc. Thesis, IFSC-USP (1998).

EMERGENT EVOLUTION OF COOPERATIVE STRUCTURES

Piero Mussio[1] and Paolo Bottoni[2]

[1]Dipartimento di Elettronica per l'Automazione
Università degli Studi di Brescia
Via Branze 38
I-25231, Brescia, Italy
[2]Dipartimento di Scienze dell'Informazione
Pictorial Computing Laboratory, Università "La Sapienza" di Roma
Via Salaria 113
I-00198 Roma, Italy

INTRODUCTION

Motivations

In many fields of science and technology, processes can be observed which cannot be properly modelled by traditional cybernetic approaches to system dynamics, as for example introduced by Bertalanffy[1] and others.[2,3] Examples include the medieval choruses of monks, the development of biological entities, the behaviour of the immune network, the life cycle of interactive software systems, the mutual adaptation of human organisations to new technologies and of technologies to human organisations.

These processes are increasingly modelled, following biological metaphors, as cooperative structures showing an emergent evolution.

By adopting this metaphor, the modeller renounces to expressing global laws in order to describe the observed process. On the contrary, the modeller focuses on local laws and communication mechanisms, and models the process as the observable behaviour of a cooperative structure. A *cooperative structure* is composed of elements, able to perform individual activities, but also able to communicate with each other, so that each single element may influence all the other elements in the structure and be influenced by them in turn.

A cooperative structure supports its elements' activities by providing a communication infrastructure. On the other hand, membership in a cooperative structure usually imposes additional constraints to the activity of the single elements, some of them deriving from the very presence of the other elements. While performing an activity, the composition of the structure, the states of the elements and the ways in which they communicate are subject to several transformations, i.e. they go through an evolution. Evolving to perform an activity, a cooperative structure exhibits a global behaviour which in general cannot be specified in

Human and Machine Perception 2: Emergence, Attention, and Creativity,
edited by Cantoni *et al.* Kluwer Academic / Plenum Publishers, New York, 1999.

advance but is unpredictable because it emerges from the interaction of the structure elements within the current situation.

Individual elements[4] may not be aware of the emergent global behaviour even if the cooperative structure as a whole may achieve some common goals. For example, the structure as a whole may survive in a hostile environment while some of its elements may not - even when a single element assumes its individual survival as one of its goals while no statement is made explicit on the population survival.

In any case, the global behaviour is perceivable by an external observer, who can a posteriori describe the evolution which emerged from the activity of the structure. In cases when it is the observer who creates the cooperative structure to model some existing process, the comparison of the observed evolution of the cooperative structure with observations on the real process can be used to validate or refuse the model.

The Point of View

The point of view assumed in this chapter is that of designers of software systems which support or perform complex processes such as image interpretation, simulation, or management of real systems. The development of such systems implies the precise, unambiguous modelling both of the complex process and of the behaviour of the system being developed.

The development of a global model is often practically unfeasible and, if human interaction is involved, also theoretically debatable.[5] Modellers increasingly renounce to global descriptions, but model the processes specifying the possible behaviours of component entities and a communication infrastructure. Designers implement these systems as cooperative structures of software elements and observe the global processes as they emerge from the evolution of the cooperative structure. A designer can adopt either a white or a black box approach in specifying the behaviour of a cooperative structure.

The *white box approach* dictates the organisation of the structure and the mechanisms by which single elements react to inputs and determine the evolution of the structure itself, as well as the description of the overall process to be performed.

The *black box approach* dictates only the description of the overall process to be performed. The structure and the mechanisms by which single elements react to inputs and determine the evolution of the structure itself is left undetermined.

The white box approach is adopted when the designer wants that the internal mechanism of the system as well its external behaviour reproduce some designed or observed processes, either real - as in the case of simulations - or hypothetical - as in some experiments in Artificial Life. If the process is real, the global behaviour of the structure as well as the local behaviours of its components can be compared with the observed behaviours of the real global and local processes to verify the consistency of the model and the structure can be used to simulate the behaviour of the real process.

The black box approach is adopted when the designer wants that the external behaviour of the system reproduce a designed or observed process, without pretending any correspondence with the internal mechanisms of the observed system.

Outline of the Chapter

The chapter discusses some perspectives opened by the use of the metaphor of cooperative structures, commenting on cases from biological modelling and pattern recognition.

We use the pattern recognition to show how a designer has the freedom to introduce models of behaviour which have no natural counterpart to solve problems which are of difficult solution by traditional techniques.

The chapter is organised as follows: first the role of the observer for emergence is illustrated; then we classify what agent communication and interaction mean in this chapter the study of this terminology allows us to discuss the use of cooperative structures to model complex systems by commenting on the example of the immune system. Then, we show that cooperation can be an emergent property and we argue on the freedom of the designer in creating agents which interact allowing complex behaviours to emerge. Finally we summarise the conclusions.

THE ROLE OF THE OBSERVER

When a process is modelled as a cooperative structure, it may be impossible to foresee its global properties which can only be described case by case by an external observer studying the structure behaviour in a specific situation. Those global properties which are easily recognisable by the external observer, but have no impact on the cooperative structure or on its elements are called *emergent.*[6]

To illustrate this point we discuss a model introduced by Lindenmayer to study the development of real biological filamentous entities.[7] To study the dependence of the development law on the structure of the observed entity, filamentous organisms are modelled as cooperative structures, whose elements are the cells of the organism. Cells are recognised as the elements which determine the evolution of the filamentous entity: however a single cell only follows its law of evolution and determines its local dynamics. Only an external observer can describe a posteriori the emerging global law of evolution of the filamentous entity as determined from the combination of the dynamics of the single cells.

Cells are modelled as having different states, each state being denoted by a different character. Each cell can change state autonomously or under the influence of some environmental input. In some particular state, and sometimes in presence of some external input, a cell can reproduce itself.

These capabilities of cells are modelled by sets of rules. The rules also reflect the communication among cells. One form of communication is pushing: when a cell reproduces itself, it pushes its neighbours, forcing them to shift away. Other forms of communication may be expressed by the rules: typically a cell can perform some state transition only if its neighbours are in some states and hence are able to influence its development. These forms of communication are modelled by context-dependent rules.

The filamentous entity itself is modelled as a string on the alphabet of cell states, as usual in Lindenmayer systems. The dynamics of this cooperative structure, the string, is discrete and emerges from the behaviour of the single elements, the cells, in that at each time step each cell undergoes a state transition, i.e. a rule is applied to it.

However, the global properties of the cooperative structure are not made explicit in the model and are only derived by the observer, as illustrated by the following example adapted from Salomaa[8] in which the development of a hypothetical type of filamentous entities is modelled.

Each filamentous specimen of this type is constituted by three kinds of cells: *developing, active* and *transient*. Cells are sensitive to light if maintained for a sufficient time (say 12 hours): light kills *transient* cells, makes *active* cells turn into *developing* ones and *developing* cells reproduce into a couple of one *active* and one *reproducing* cells. In a dark period, cells exhibit a different behaviour: *active* cells reproduce into couples of one *active* and one *transient* cells, *developing* cells reproduce into couples of one active and one *transient* cells and *transient* cells reproduce into triples of *transient* cells. Let us assume that symbols in the alphabet A={D,A,T} denote developing, active and transient cells respectively and describe the local law of evolution of the three types of cells by the following rules:

Light: $A \rightarrow D, D \rightarrow AD, T \rightarrow \lambda$

Dark: $A \rightarrow AT, D \rightarrow AT, T \rightarrow TTT$

where λ denotes an empty string.

A specimen is described as the sequence of letters which describe the cells composing it: TTAAAD describes a filamentous entity -a *population*- of 6 cells. To describe the development occurring during one period of light (or dark) we apply the light (dark) rules by replacing every occurrence of every letter by the right-hand side of the rule for the letter.

For example TTAAAD becomes DDDAD after one period of light, and we write $TTAAAD \Rightarrow^{L} DDDAD$ to indicate this replacement. If alternate dark and light periods then follow, the specimen exhibits the following emerging behaviour:

$DDDAD \Rightarrow^{D} ATATATATAT=(AT)^5 \Rightarrow^{L} DDDDD=(D)^5 \Rightarrow^{D} (AT)^5 \Rightarrow^{L} (D)^5 \Rightarrow^{D} (AT)^5$etc..

An observer may note that the specimen changes during the night, while it reverts to the previous state during the day, and recognise an occurrence of what Salomaa calls stagnation.

This behaviour is independent of the specific specimen: it can be demonstrated that whichever the initial configuration, if one dark and one light periods are alternated, a specimen of this type either disappears or stagnation occurs within two steps.

The behaviour of a specimen depends on the environmental input: if the specimen is subject to a different regime of light and dark period, it develops with different laws.

For example, if the specimen is maintained in eternal dark it becomes malignant: that is if we denote by w_i the string denoting the state of a specimen at step i and with $|w_i|$ the number of letters composing it, there is no polynomial d(i) such that $|w_i| \leq d(i)$ for all $i=1,2,3,\ldots$

Salomaa comments that in this situations these specimens 'face what is often called prognosis pessima in medicine'.

With these local laws, the development of the cooperative structure, the specimen, is linear in time, whichever the regime of illumination one assumes. Some experimental observations of real filamentous entities showed that some types of entities develop following different laws, e.g. logarithmic. Lindenmayer demonstrated that only context-dependent rules allow the description of a logarithmic development. From the biological point of view this means that a logarithmic development of a filamentous organism can only occur if a cell development depends also on the action of its neighbour cells and not only on its genetic structure and environmental inputs.

Some observations are in order: the modeller had to model a case of evolution, the development of a living entity. The living entity was modelled within the white box approach as a cooperative structure composed by cells.

The specimen development is explained by establishing the rules by which the component cells react to inputs - the light or dark quanta. Cells only obey to their local rules and, in an uncontrolled environment, the modeller cannot compute the system development before looking at the input. The behaviour of the specimen emerges as the result of the local actions of the cells.

Cells are modelled with the black box approach: only the description of their overall behaviour is given by the rules without giving any insight on their internal mechanisms.

This is a very general point: every time the white box approach is assumed, the modeller has to choose an atomic level at which the component elements are described as a black box.[9]

AGENTS, COMMUNICATION AND INTERACTION

In Lindenmayer models, the elements of the cooperative structures are the cells, recognised by the modeller as entities which develop their own processes so that they are capable to exhibit a specific behaviour. In this case, cells are modelled as agents.

By *agent* we identify every entity which is recognised by an observer as a whole entity, irrespective of its nature, whether biological, human, software, hardware.[10] An agent is able to perform an *action*, i.e. to bring to bear changes in its environment or in the state of some other agent.

The actions an agent performs may depend on stimuli (inputs) from the environment, but also on the agent's capabilities and resources, so that the same stimulus can produce different responses at different times.

This dependence of an agent's response on some internal property is made precise introducing the systemic *state* variable, whose value describes the current agent's capabilities and resources.[2,9]

The environment appears to the agent as the support from which messages arrive, both explicit and implicit. Messages are vehicled by some mediators which an agent may explicitly sense and perceive. In this case we say that the message is *explicit*.

Messages are *implicit* when the agent senses but does not perceive the mediator: for example a first agent is infected by a second one. The first agent is not aware of being infected but looses some (even only potential) capabilities.

Note that the environment seen by an observer contains a third kind of messages: those which are neither sensed nor perceived by the agent.

Sensing messages does not imply any action of the agent toward the environment. It implies a state change of the agent, which may involve gaining or loosing the capability of reacting to some class of messages. If an action toward the environment occurs, we say that the agent *reacts* to the messages.

The set of messages an agent is able to sense is called the *input set* of the agent. The environment of an agent can be constituted by another specific agent, by a set of agents or by the whole universe of agents.

Therefore, agents communicate and interact with other agents. Communication amounts to the manifestation of some sort of influence (a modification of the state of) an agent has upon another agent. In general, communication is vehicled by some mediators, possibly exploiting some substrate.

The abstract notion of *interaction* is taken to indicate the existence of some mutual relation among agents such that transitions in the state of an agent depend upon a transitions in the state of the other.[11] Such an interaction can be explicit or implicit. Explicit forms of interaction are realised through some synchronisation mechanism, as in protocols based on handshaking in hardware and software,[12] or in the aggression of a virus against cells in biology, or in the physical impact of a body onto another.

Implicit forms of interaction are realised through mutual dependencies which can appear either instantaneous to some external observer, as in the case of interaction between a particle and a field, or deferred, as in the case of the dynamics of populations of predators and preys, or of the debate on a newsgroup.

Actually, whether an interaction can be deemed as implicit or explicit is a matter of the level of abstraction at which a phenomenon is observed or modelled. A typical case is that of gravity, which can be regarded as the consequence of distortions in the space-time continuum induced by the presence of bodies, or the consequence of the exchange of specific mediators.

Hence, we view the notion of *communication* as defining the act (or sequence of acts) through which interaction is established, rather than restricting it to design only explicit interactions. In an agent-oriented perspective, we may define communication as the common participation of two agents to complementary activities or to a same event.[13-15] For communication to give rise to interaction, it is necessary that the recipient agents be endowed with some "sensory" apparatus able to convert the mediator into some modification of its internal state.

Note that in this definition, neither communication nor interaction implies a form of intentionality, either in the emitter or in the receiver: any modification of the environment can constitute a communication.

For example, each modification of the spatial configuration of the membrane of a cell produces the release of chemical mediators. A chemical gradient is thus produced in the surrounding space and constitutes the stimulus to some other cells to direct their movement towards (or away from) the source of the mediators. The speed with which another cell can direct towards the source is in any case determined also by other factors, e.g. the resistance of the medium through which cells move and the cell energy, i.e. its state.

Analogous mechanisms seem to determine the establishing of proximity relations among humans, where not only chemical aspects are involved, but also visible signs, such as the posture of the body, or audible ones, such as the voice tone. The reaction to these stimuli again depends on the state of the agents, but now this also incudes cultural aspects, even determined by the social environment in which the agents are embedded.

In all these cases interaction occurs through a common cyclic pattern of type:

- diffusion of signals by some agents;
- perception of signals realised by some apparatus of another agent;
- reaction according to the state and the environmental constraints, producing new signals.

In general, an agent is able to perceive only a part of its environment. Perception can be limited in space (e.g. an agent in a given state is able to perceive only up to a given distance, or only stimuli coming from a given direction) and/or in quality (e.g. an agent in a given state is able to perceive only within a certain range of wavelengths, or only above some intensity threshold).[16] The perceptual capability of an agent can vary in time, due for example to saturation or ageing phenomena.

THE WHITE BOX: AGENTS AS COOPERATIVE STRUCTURES

In this section the white box approach is discussed by modelling the immune system behaviour under a bacterial attack. It is assumed that the behaviour of a biological entity is modelled by the behaviour of a set of agents. The state of each agent is determined by the available resources, described as a finite set of biological attributes, and the agent dynamics is determined by the composition of biological functions from a finite set.

In this approach, the most elementary units of investigation are not the biological entities, but the functions they perform and the variables affected by these functions. Hence, biological entities are modelled as agents, which are cooperative structures typed according to their repertoire of functions and variables. Agents can be derived by composing subsets of elements -the *repertoire*- into a cooperative structure. In turn agents interact to pursue either partaken or competing goals and become organised into cooperative structures called *populations*. The whole immune system is modelled as a cooperative structure in which different populations interact determining the emergence of its global behaviour.

To verify the model being developed against experimental data, each biological attribute is named, associated with an attribute variable taking values on a set of symbols or numbers according to the scale in which the attribute is measured and denoted by icons framed by a circle (see Table 1a). Each biological function is named, modelled as a mathematical function and denoted by an icon framed by a rectangle (Table 1b). Agents are named, modelled as sets of attribute variables and mathematical functions and denoted by icons framed in a hexagon (Table 2). Three types of mathematical functions have been envisaged: next state, transition and output functions.

Next state functions can map either a single variable or a pair of attribute variables into a single variable. Functions on single variables model an agent evolution only using its internal resources. Functions on pairs model an agent evolution using both internal resources and resources captured in the interaction with its environment.

Table 1. A subset of the set of elements modelling the immune system. a) A subset of the set of Characteristic Attributes. b) A subset of the set of elementary next state, interaction, creation and transition functions.

ATTRIBUTES	DEFINITION RANGE	ICON
Antigen peptides	Set of character triples	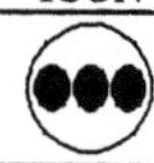
Opsonisation by antibody	{0,1}	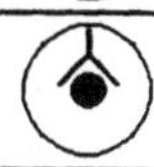

a)

NEXT STATE FUNCTION	ICON
Ageing	
Capability of specific uptake	
OUTPUT FUNCTION	ICON
Proliferate into	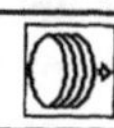
TRANSITION FUNCTION	ICON
if time out, **then**	
if specific uptaking occurs, **then**	

b)

Transition functions map pairs formed by a set of attribute variables and a function into a different such pair. The result of the function is denoted by an agent possessing the required set of attribute variables. They describe how an agent evolves into a different agent using its internal resources or because of an external event. In the simulation, this results into making the relevant agent active. Two kinds of transition can occur: endogenous and exogenous.

In endogenous transitions, some event internal to an agent, describing a biological entity able to perform a set of elementary functions, causes its transformation into a different agent describing the same entity and able to perform a different set of elementary functions. A typical transformation is death: the agent becomes the null agent, not able of any function. Death can be endogenous, when the agent dies because the time assigned to it ends. Death can be exogenous when the agent dies because uptaken by another one.

Output functions map pairs from the Cartesian product of the set of attribute variables and of set of agents into the set of agents and describe how agents proliferate or output new agents (see Table 1b).

Agents are described at a high level of abstraction by their icons and at a lower level

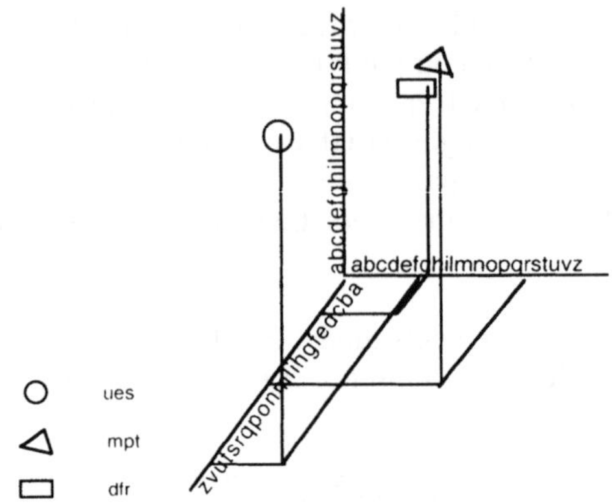

Figure 1. A representation of a population of Bacteria.

by enumerating the attribute variables which characterise its state and the mathematical functions modelling their capabilities (Table 2).

The interaction among agents

The biological entity Bacterium is characterised in this model by one resource -i.e. the antigenic peptides present on its membrane -and in some situations by a second one- an antibody which is bound to its membrane (in biological terms the Bacterium is opsonised). The state of an agent modelling a Bacterium is defined by enumerating the types of peptides and antibodies present on its membrane. Each peptide type is denoted by a triple of alphabetic characters, the state variable of a Bacterium as a set of triples. When a Bacterium proliferates, it generates a set of Bacteria, some characterised by the same triple as the parent and some by different ones, according to a mutation law.

We characterise a population of Bacteria in a 3-D space whose axis are labelled by characters in alphabetic order. Every triple of peptides is represented by an icon (Figure 1). If one or more Bacteria exist in the population characterised by this triple of peptides, the icon is positioned at the point whose coordinates correspond to its 3 characters. The size of the icon is proportional to the number of Bacteria exhibiting this peptide present in the population.

The Bacterium dynamics is determined by its ability of performing the following biological functions: a) to reproduce itself - i.e. to perform a biological function of outputting new specimens of its same biological type; b) to be bound by an antibody - which changes its capability of being captured; c) to be uptaken by other entities -and die.

For example, Figure 2 sketches the development of an isolated population of Bacteria. An isolated Bacterium is involved only in the activity of reproduction, during which small mutations can occur originating new specimens characterised by different triples of peptides. These new peptides are represented by deformations of the original shape.

B-lymphocytes go through a similar evolution. However they obtain resources for reproduction from uptaking Bacteria. Each B-lymphocyte is characterised by a receptor, whose shape matches some peptides which may appear on the Bacterium membrane. Only if the receptor shape matches the peptide type can the lymphocyte uptake the Bacterium.

The interactions among B-lymphocytes and Bacteria can be modelled in the following way. When a B-lymphocyte meets a Bacterium, if the B-lymphocyte recognises the Bacterium, it uptakes the Bacterium and exposes some of the Bacterium components (the peptides) on its surface becoming available to interaction for proliferation.

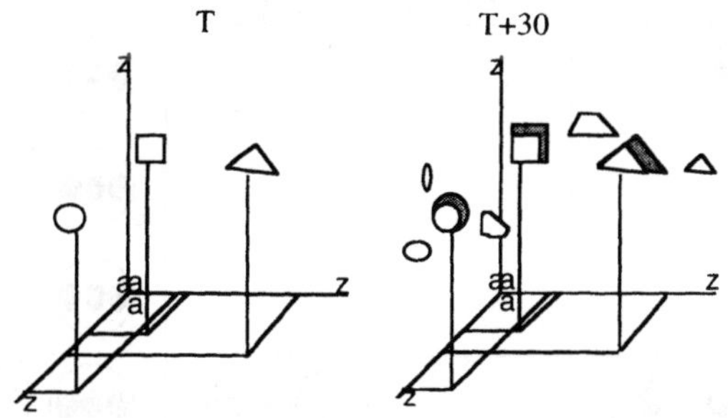

Figure 2. Two time steps in a sketch of the development of a population of isolated Bacteria. a) Initial composition. b) The composition of the population after 30 time steps of evolution.

If the Bacterium is originally described by a Bact agent and the B-lymphocyte by an LB agent this story is told as:

a) the Bact agent state is characterised by a set of peptides;

b) if the LB agent recognises the Bact agent, then the story goes on, otherwise they ignore each other;

c) the LB agent uptakes the Bact agent and the Bact agent dies;

d) a transition function of the LB agent transforms it into an LB-APC agent exposing the peptides captured and able of different interaction functions from the original LB agent.

Note that in this interaction the biological entity Bacterium can be described by a single agent and then disappears; the biological entity B-lymphocyte is described by two agents: the first mimicking its activities when it has not yet uptaken any Bacteria. The second mimicking its activities when it has eaten at least a Bacterium, so that it is able to exhibit the peptides and go through different interactions which allow it to proliferate.

However, even in such a simple experiment, the biological entity Bacterium cannot be described by only one agent, if we take into account the existence of Macrophages. In fact, a Bacterium can be opsonised, i.e. the Bacterium meets an antibody or a complement factor and this entity attaches itself on the Bacterium surface.

The Bacterium does not perceive this change. However, Macrophages in the system do. These are cells which uptake any entity alien to the organism; but they can recognise an opsonised Bacterium both because of its peptides and because of the bound antibody. Therefore a Macrophage can uptake a bare Bacterium with only one attack strategy and an opsonised Bacterium with two different strategies. The probability to uptake an opsonised Bacterium is higher than for a bare Bacterium.

Hence two agents are required to model the Bacterium behaviour: a bare Bacterium agent (bb) and an opsonised Bacterium agent (ob). A bb is characterised by its peptides - iconified as a circle containing 3 black spots. The bb is not able to interact with other agents; it can proliferate producing one copy of itself on average every 30 minutes. When opsonised, a bb becomes an opsonised Bacterium agent (ob). In Table 2, this transition is iconified in the first line of the transition column for bb. The other transitions express that when uptaken by a Macrophage, a bb dies (second line); and that when uptaken by a B–lymphocyte it also dies (third line).

The ob agent differs from the bb in that it is opsonised - i.e. it has one more attribute; it can be reopsonised but this does not change its behaviour; it can be more easily uptaken by Macrophage agents because of the bound antibody, as iconified in the fourth line in the transition column for ob.

Table 2. The set of two agents modelling the behaviour of a Bacterium.

AGENT	ICON	ATTRIBUTE	NS & INT. FUNCTION	CREATION FUNCTION	TRANSITION
Idle Bacterium				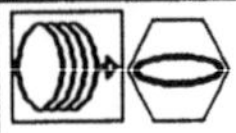	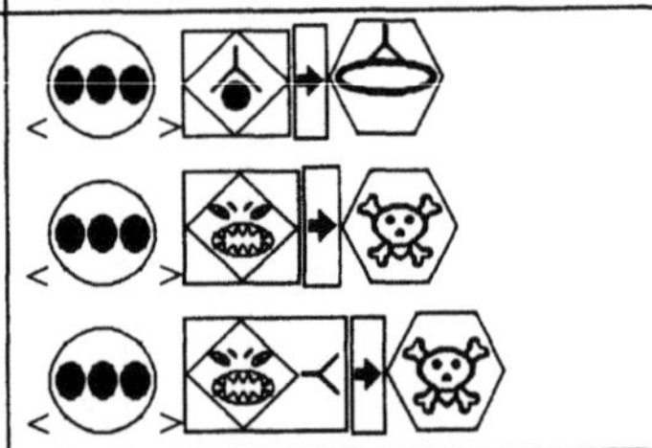
Ab-opsonised Bacterium		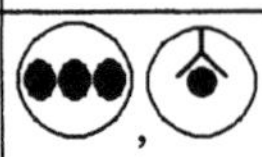		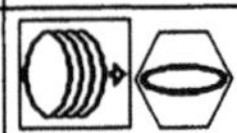	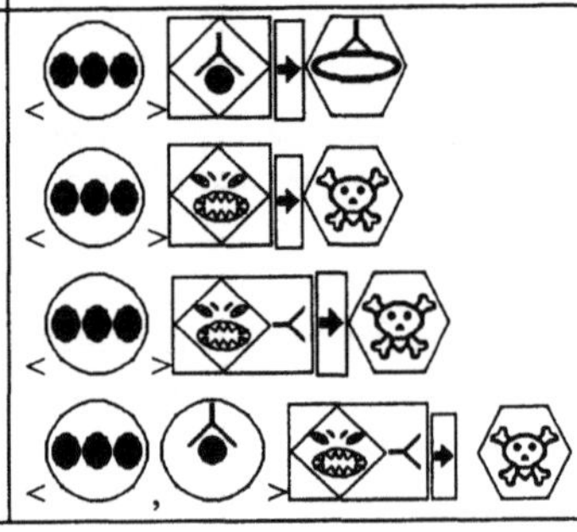

Emergence and cooperation from proliferation and interaction

The goals pursued by an agent can be made explicit (in the agent description a variable states them, as in classical cybernetic modelling) or implicit (they only emerge from the agent or population activity).

Agents pursue their implicit or explicit goals by exchanging messages or modifying some common memory support. Individual agents may compete for resources, yet produce a common effect. Populations pursue implicit goals by the same means, but also using reproduction mechanisms.

The immune system model is an example of a population formed by two populations, the population of Macrophages and of that of B-lymphocytes. It has the implicit goal of limiting the diffusion of Bacteria, i.e. of contrasting a different population. This goal is not stated in any variable, but emerges from the immune system dynamics.

The immune system improves its ability of contrasting Bacteria diffusion by a proliferation mechanism – the somatic mutation.

The population of Bacteria improves its ability of escaping from B-lymphocytes and surviving by a proliferation mechanism – escape mutants.

In detail, when a population of Bacteria meets one of B-lymphocytes, interactions between single B-lymphocytes and single Bacteria occur; Bacteria whose peptides are recognisable by the lymphocytes receptors are uptaken and tend to disappear. B–lymphocytes which uptake Bacteria become able to reproduce, generating offsprings which can in turn uptake Bacteria. However, the two populations are subject to mutations. Bacteria generated with a set of peptides partially matching the receptors of the existing B–lymphocytes have a higher probability of surviving and reproduce. B-lymphocytes generated with a receptor better matching existing peptides have a higher probability to reproduce.

This situation is sketched in Figure 3 in which the somatic mutations of the B–lymphocytes allow the destruction of the invading Bacteria and result in a B–lymphocyte population of different characteristics. The pool of receptors owned by the

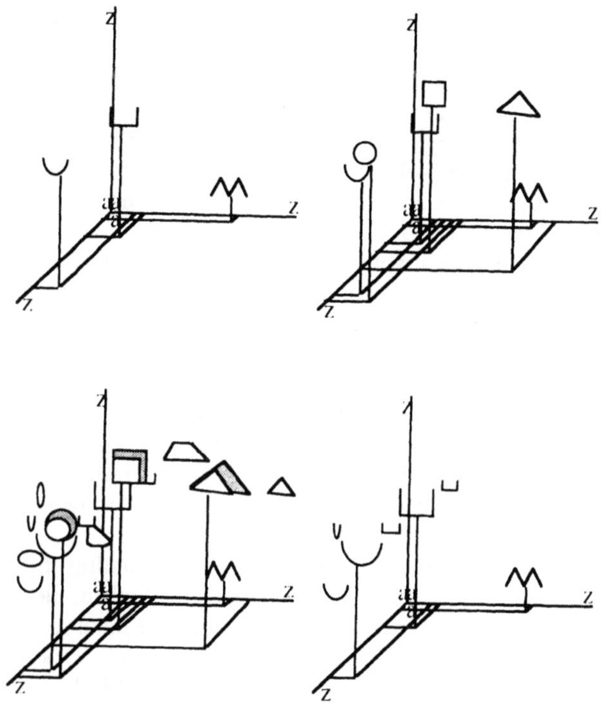

Figure 3. Populations of Bacteria and receptors interact.

population of B-lymphocytes represents the identity of the immune system, in that it identifies the types of Bacteria whose diffusion can be contrasted. The somatic mutation derived by the interaction with a population of Bacteria leads to the emergence of a new identity of the immune system.

To describe this process, a set of reactive cooperative structures (agents) whose state variable is defined by the Cartesian product of the set of attributes values and whose dynamics emerges from the composition of a set of next state or output functions have been defined. The structure of agents in this model follows the "Steels style"[17] in that their action results from the composition of the effects obtained by the execution of each single algorithm in their repertoire. This contrasts with the "Brooks style"[18] in which a subsumptive structure is imposed on the functions to enable or inhibit the activation of specific functions.

The model can be implemented by several techniques. If the algorithms computing the defined functions are developed in software and associated with icons, following the methods described by the authors elsewhere,[19] a visual language can be defined and used to program simulations.

SOMETIMES COOPERATION IS AN EMERGENT PROPERTY

There are situations in which cooperation among elements of a structure is itself an emergent property. These cases take place when the observer looking at a structure whose elements are passive, attributes some capability to them and sees some new patterns emerge because of this capability.

A typical example occurs when experts in some field communicate and reason through multimedia documents. Experts generally develop a notation to express and support their reasoning and to describe the intermediate and final results of these activities. They use this notation to produce documents in which texts, pictures and graphs are organised and assume a meaning according to the conventions adopted by them. The documents appear as sets of *characteristic structures* (**cs**), which constitute functional or perceptual units for an expert reading or composing a document. In the textual part of the document, characters of different fonts are **css**, as well as the words formed by them. In the pictorial part, drawings, graphs, circles, lines and dots are organised to represent the entities of interest. Experts name each type of **cs** they use and characterise it by a set of attributes, each attribute assuming its value in a well defined set of values. The name of the type and the set of values assumed by the attributes identify an instance structure of a given type and constitute its *description*. We call the association of a **cs** and its description a *characteristic pattern* (**cp**).

An expert who needs to communicate some concepts composes the document by assembling simpler **cps** into more complex ones following the established conventions, so that the document itself becomes a complex **cp**. However, a different expert looking at the document sees a complex **cs**, which can be correctly interpreted only knowing the set of characteristic patterns defined by the first expert and the conventions used to compose them.

Often, two experts look at a document, aimed at reaching different goals. In these cases, it may occur that they see different structures emerge from the document at hand.

For example, a production engineer may look at the technical drawing of Figure 4, which was produced by a designer. This drawing represents five **css** each one representing for an engineer a well defined **cp** - the top view of a threaded hole.[20]

The designer has placed them, taking into account the mechanical properties of the part to be produced. But the production engineer reasons on the document to derive the better path an instrument has to follow to produce the mechanical part. In his/her mind the structures in the drawing cooperate to let a pattern which is not traced in the image emerge as shown in Figure 5 b) and c).

This process occurs frequently when different experts reason on scientific or technical documents: the patterns which are recognised and those which emerge in the document depend on the observer's culture. Emergence requires both a space in which the phenomena of interest can be represented and can be observed and a hierarchy of descriptions of the phenomena of interest – e.g. the path emergent from the composition of the top views of threaded holes.

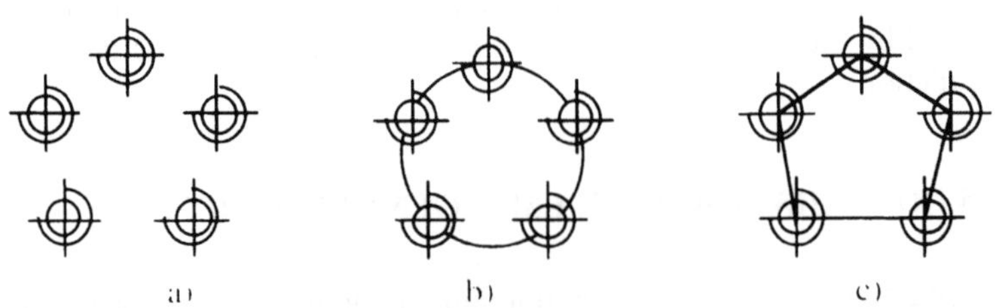

Figure 4. a) A technical drawing in which five top views of threaded hole appear; b) a first possible pattern organising the five structures; c) a second possible pattern.

THE BLACK BOX: DESIGNING ON EMERGENCE

Designers are often required only to implement artefacts which perform some tasks but are not restricted to reproduce some internal mechanism of natural systems. They become free to develop systems which have no counterpart in the biological reality even when they exploit biologically inspired mechanisms. For example, they can use emergence in cooperative structures as a paradigm to develop populations of agents which exhibit complex global behaviours, at the only cost of defining simple local "rules" which determine the behaviour of agents and the communication infrastructure. The designers are free to attribute to agents cognitive abilities which are characteristics of the observers, so that coordination emerges from agent interaction, or to introduce ad hoc reproduction mechanisms, so that a population emerges which is adapted to the solution of the problem at hand.

In this way, the agents are endowed with new communication and computational abilities which no non-human biological entity in the real world possesses and some global patterns of behaviour -which in the real world emerge due to the observer's cognitive abilities- emerge in the virtual world as an evolution of the cooperative structure.

Two examples from pattern recognition illustrate these possibilities. In both cases cooperative structures are designed which obtain image descriptions by identifying the characteristic structures in the images.

In the first case, the cooperative structure incorporates the observer's ability of tuning observational tools to the different characteristics by a specialised system of reproduction. Agents generate their offspring adapting them to the data to be interpreted. In the second, the elements communicate among themselves organising a network, which describes the set of data at hand.

In the first case, a population of agents develops from an ancestor adapting to meet the environment characteristics and a description of an image emerges from their cooperation. The environment is represented by a digital image to be analysed, whose size and content are not predefined. The goals to be achieved are characteristic patterns defined by sets of rules, so that the agent using them is able not to recognise an instance, but a (possibly

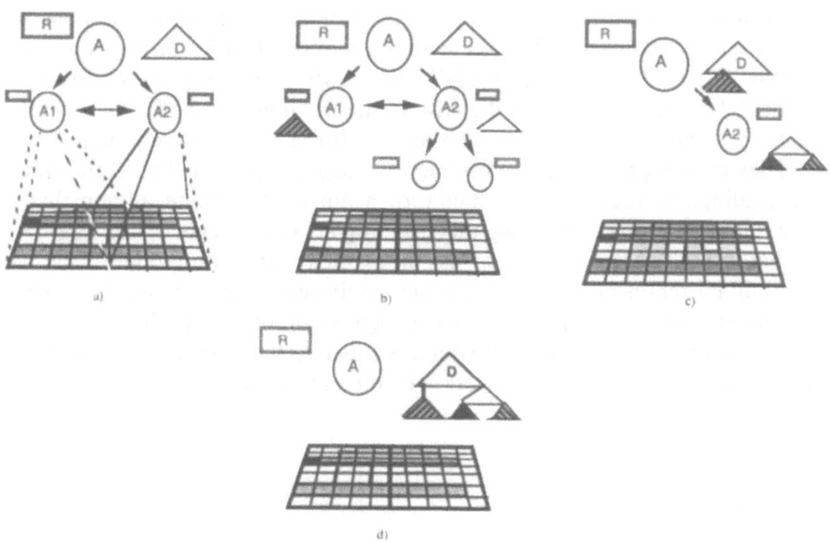

Figure 5. The ancestor A generates a population of customised agents to interpret an image. a) A generates two offsprings A1 and A2; b) A1 achieves its goals, while A2 generates two more specialised executors; c) A1 ceases to exist, while A2 collects results from its offsprings; d) the ancestor A collects the results produced by the generated agents, which disappear.

infinite) set of instances of characteristic structures, i.e. a pictorial language.[21] Given a set of data, an agent examines them and generates a description in which the specific instances of characteristic patterns recognised in the data are enumerated.

According to the set of rules available to the agent, two types of agents exist: those (called *managers*) which examine the image or a description of the results obtained so far to generate one or more specialised agents, and those able either to process images or to analyse them obtaining a description (called *executors*).[22]

The process is sketched in Figure 5, in the case of an imge representing a hut. The ancestor agent A is a manager which applies its set of rules R to interpret the image, having an initial description D available. A identifies two homogeneous regions in the image, and generates an executor A1 and a manager A2. A1 is endowed with a customised set of rules and is in charge of describing the first region; A2 is endowed with a second customised set of rules to generate strategies for the description of the second region (Figure 5a). A1 succeeds in describing its region and generates a second description (dark triangle in Figure 5b), while A2 needs to generate two more executors (Figure 5b). A1 reports its results to A and ceases to exist, so that A may merge the new description with D, while A2 receives reports from its two executors (Figure 5c). Then A2 reports its results to A and ceases to exist, and A merges all the received descriptions in the final description (Figure 5d).

Note that the number, the type and the characteristics of the offsprings (i.e. the set of rules they are endowed with) cannot be foreseen a priori, but emerge from the interaction of the parent with the image and the descriptions obtained so far. Generation is not a mutation, but a differentiation from the parent, so that it does not have a biological counterpart. The size and the composition of the population emerges from the interaction with the image. Given an image, the number of generated elements is always finite, but since there is no predefined limit to an image size, there is no upper limit to the number of agents which can be generated. Some results from this type of approach are described in works by the authors.[22,24]

In a second case, a fixed population of agents exist, each agent being able to recognise a simple or complex family of characteristic structures. Agents cooperate to construct the description of the whole image, disambiguating the meaning of those characteristic structures which can have multiple meanings.[25] For example, let us suppose to have a population of specialised agents able to classify sets of pixels into simple structures such as "roof", "base", "road", "wheel", "body" and "wall", or complex structures such as "house", "hut", "car" and "bike". In the case of the hut image, "roof", "base", "road", and "wall" identify some sets of pixels which are candidate to be characteristic structures of that type. They exchange messages with the other agents, until the hut agent organises the individual findings to identify an instance of a hut in the image. Given this complex recognition, interpretations contrasting this meaning are discharged.[26]

Note that in both cases, agents possess an explicit, symbolic model of the world that they use to build a representation of the actual environment in which they exist. They are therefore deliberative agents. In the first case agents deliberate on how to generate new agents: a population of agents emerges which is suited to the interpretation of the images at hand, while in the second case the network which represents the image description emerges as a result of agent deliberation.

CONCLUSIONS

Emergence in cooperative structures is a new modelling paradigm suggested from the study of biological processes, whose behaviour cannot be described by traditional mechanisms with central, explicit control. The chapter discusses this paradigm from the point of view of system designers, illustrating a sequence of biological models proposed to explain processes of increasing complexity. The aim is that of illustrating how cooperation of simple structures determines the emergence of complex behaviours and how evolution may lead to the forming of structures better suited to perform the process under study. These simple, structures can be implemented as software or hardware agents. These agents can exhibit different types of behaviour, from pure emergent functionality (à la Brooks: an individual as a set of interacting entities), to deliberative (an individual as a whole having explicit knowledge, goals and plan, i.e. represented but variables which can be instantiated in reasoning processes, in making decisions, or as rules which need to be selected by pattern matching).[6,18] The role of the human observers in describing the behaviour of the cooperative structure is studied and their limits in foreseeing it are illustrated. It is also argued that the observer becoming the designer of artefacts can evolve this paradigm proposing models which have no counterparts in the biological world. In this way systems can be designed which exhibit a behaviour dependent on the environment in which they perform their tasks.

REFERENCES

1. L.van Bertalanffy, *General Systems Theory*, Breziller, New York (1969).
2. M.D. Mesarovic and Y. Takahara, *Abstract Systems Theory*, Springer, Berlin (1989).
3. G. Trautteur, Distinction and reflection, in: *Consciousness: Distinction and reflection*, G. Trautteur ed., Bibliopolis, Napoli, (1995).
4. H.D. Buckhard, How to define agent properties or: what is a fair agent?, in: *From reaction cognition*, C. Castelfranchi and J.P. Muller eds., Springer, Berlin (1995).
5. P. Wegner, Why interaction is more powerful than algorithms, *Communications of the ACM*, 40(7):80, (1997).
6. C. Castelfranchi and J.P. Muller, Introduction, in: *From reaction cognition*, C. Castelfranchi and J.P. Muller eds., Springer, Berlin (1995).
7. A. Linndenmayer, Mathematical models for cellular interaction in development I and II, *Journal of Theoretical Biology*, 18:280, (1968).
8. A. Salomaa, Stagnation and malignancy: Growth patterns in artificial life, in: *Artificial Life: Grammatical Models*, Gh. Paun ed., Black Sea University Press, Bucharest (1995).
9. P. Mussio, Exploring living systems: the state space approach in: *Proceedings of the First International Conference The Future of Science Has Begun*, (1993).
10. G. Kiss, Agent dynamics, in: *Foundations of Distributed Artificial Intelligence*, G.M.P. O'Hare and N.R. Jennings eds., Wiley, New York (1996).
11. R. Milner, Computing is interaction, in: *Proc. IFIP 13th World Computer Congress. Vol.1: Technology and Foundations*, B.Pehrson and I.Simon eds., Elsevier, Amsterdam (1994).
12. N. Gehani, *Ada concurrent programming*, Prentice Hall, Englewood Cliffs, (1984).
13. C.A.R. Hoare, *Communicating Sequential Processes*, Prentice-Hall, Englewood Cliffs, (1985).
14. R. Milner, *Communication and Concurrency*, Prentice Hall, Englewood Cliffs, (1989).
15. M. Snoeck and G. Dedene, Existence dependency: the key to semantic integrity between structural and behavioral aspects of object types, *IEEE Transactions on Software Engineering*, 24:233, (1998).
16. J. Kelemen, Artificial life: Describing life-like behaviours in computational frameworks, in: *Artificial Life: Grammatical Models*, Gh. Paun ed., Black Sea University Press, Bucharest (1995).
17. L. Steels and M. Tokoro, Artificial life and real world computing, in: *Computer Science Today*, J.v. Leeuwen ed., Springer, Berlin (1995).
18. R.A. Brooks, Intelligence without representation, *Artificial Intelligence*, 47:139 (1991).
19. P. Bottoni, M.F. Costabile, S. Levialdi, and P. Mussio, Specifying dialog control in visual interactive systems, *Journal of Visual Languages and Computing*, 9:535 (1998).
20. U. Cugini, P. Ferri, P. Mussio, and M. Protti, Pattern-directed restoration and vectorization of digitized engineering drawings, *Computer & Graphics*, 8:337 (1984).
21. A. Rosenfeld and R.Siromoney, Pictorial languages, *Languages of Design*, 1:229 (1993).
22. N. Bianchi, P. Bottoni, C. Spinu, C. Garbay, and P. Mussio, Distributed plan construction and execution for medical image interpretation, in *Proc. AIME'97*, Springer, Berlin (1997).
23. C. Spinu, C. Garbay, and J.M. Chassery, A cooperative and adaptive approach to medical image segmentation, in: *Artificial Intelligence in Medicine*, P. Barahona, M. Stefanelli and J. Wyatt eds., Springer, Berlin (1995).

24. N. Bianchi, P. Bottoni, C. Spinu, C. Garbay, and P. Mussio, Situated image understanding in a multi-agent framework, to appear in: *International Journal of Pattern Recognition and Artificial Intelligence*, (1998).
25. P. Mussio, On positive use of ambiguity in machine vision and image understanding, in: *Human and Machine Vision. Analogies and Divergencies*, V. Cantoni ed., Plenum Press, New York (1994).
26. S. Dance, T. Caelli, and Z.-Q. Liu, A concurrent, hierarchical approach to symbolic dynamic scene interpretation, *Pattern Recognition*, 29:1891 (1996).

INTERVENTION IN EVOLUTION: (UN)PREDICTABILITY VERSUS CONTROL

Jörg D. Becker

Universität der Bundeswehr München, ET/9, D-85577 Neubiberg
Starnberg/Munich
e-mail: Joerg.Becker@unibw-muenchen.de
http://home.t-online.de/home/ICAS-e.V./homepage.htm

ABSTRACT

Evolutionary games belong to the most fascinating examples of learning systems. The notion of learning is formalised by introducing a rigorous definition of pragmatic information as the precision to which constructs and ideas fulfil a given vision. We then discuss various aspects of encouraging creativity and innovation in evolutionary systems.

INTRODUCTION

Several years ago, Umberto Eco came to Munich to present his book "Il nome della rosa".[1] After his presentation he was asked why, being a scientist, he had written a novel. He replied that he had noticed that people would not understand his theories, and therefore he decided to reveal them in a novel.

I found myself in a similar situation; so I decided to write a new version of an ancient play: Antigone.[2]

When Oedipus dies he leaves four children:
Eteocles (who should have got the crown after Oedipus' death);
Polyneices (who fights with Eteocles for the crown);
Antigone (who wants to prohibit the fight, but she comes too late - both brothers are killed in the fight);
and Ismene (who is a victim of reengineering in this version of the play).
Finally, there is Haemon (who is in love with Antigone which makes the situation still more complicated), and there is Creon (Haemon's father and, at the same time, Antigone's uncle).
This is the situation in which the play starts.

Act I: Creon crowns himself and declares that nobody is allowed to bury Polyneices. (The play is rather short and thus suits an experimental theatre.)

Act II: Antigone finds the corpse of Polyneices. (Polyneices is a thankless role but nevertheless quite adequate for a novice.) Antigone starts to mourn -
"*To bury or not to bury - that's the question...*"
because if she would not bury her brother he would (according to the Greek traditions) not find peace; but if she buried him Creon would wall her in, and she would have to die. At last she takes the spade to dig the grave for her brother ... when she suddenly remembers what Hegel said:
"*To bury AND NOT to bury - that's the answer!*"
Antigone builds a mausoleum around the corpse of her brother. (This can easily be done on the stage using cardboard boxes or styrofoam blocks).

Act III: Sarastro enters the stage - (you might say: What's he doing here? ... But we must leave some work to the critics!) -, crowns Antigone, praises the victory of feminine logics over masculine logics (actually, feminine logics is rather related to dialectics: truth is what is useful for a woman), praises the victory of the THREE over the TWO, and marries Antigone to Haemon.

Thus, Sophocles' tragedy has been transformed into a parable. This demonstrates the progress in philosophy since the ancient times.

GENERAL MODEL OF A LEARNING SYSTEM

Our new Antigone has shown creativity and innovation to solve her complex problem in a constructive way. In order to model such processes formally we recall the general model of a learning system[3] (see Fig.1).

A robot creates constructs from a set of (given or found) building blocks and construction rules, in some environment. The constructs are evaluated according to the vision of the theologian. If among the construction rules there are also learning rules the system may be able to learn. In order to define learning as a gain of information we have to find a quantitative measure.

Formally we assume that the space of all possible constructs is equiped with suitable mathematical structures, in particular, operations which we call $\{+,-\}$ denoting composition or decomposition, and a norm $\|\psi\|$ which measures the size of a construct ψ. The vision is represented by a vision operator V which acts as an analysis (i.e. decomposition) operator. If V happens to decompose a construct ψ completely, $V\,\psi = 0$, we call ψ a solution. There may be one, several, many, or none such ψ.

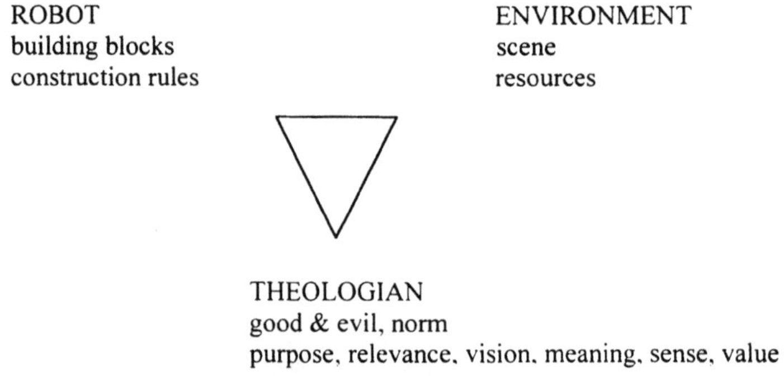

Figure 1. General Model of a Learning System.

Hartley originally defined information as the precision to which we can locate a given object in a given search space. We can generalise this concept by defining pragmatic information as the precision to which the vision of the theologian is fulfilled by the construct of the robot:

$$S := -\log \|V\psi\| \qquad (1)$$

This happens to be related to Misra's entropy concept which may be written as

$$S = \log \|\Lambda\rho\| \qquad (2)$$

ρ being the density matrix, and Λ the "square root" of the (Misra) entropy production operator. Λ may be interpreted as the vision about the predictabililty of the system.[4]

We note that pragmatic information cannot be localised in any of the instances of the model - information is distributed; and that information is subjective because it depends on the vision of the theologian. We also note that there is no learning without a vision, i. e. without reference to a value.

We quote three examples:

• the vision "reproduction of a given construct ψ_g":

$$V\psi = \psi - \psi_g\ ;\ S = -\log\|\psi - \psi_g\| \qquad (3)$$

• the ground state ψ_0 of the harmonic oscillator may be characterised by the "annihilation operator" a (the "vision operator" of the ground state):

$$a\,\psi_0 = 0 \qquad (4)$$

In the Schrödinger representation we have $a = i\,(q + d/dq)$, acting on wave functions $\psi(q)$; q the (dimensionless) position variable; which yields the differential equation

$$i(q + \frac{d}{dq})\psi_0(q) = 0 \qquad (5)$$

• the vision "construct any mirror symmetric form" may be formalised by

$$V = 1 - P \qquad (6)$$

being P the mirror operator with respect to a given axis.

We may call this the palindrome vision operator.

EVOLUTION – A CREATIVE WAY OF SOLVING PROBLEMS

An outstanding mechanism for autonomous learning is evolution, or rather breeding. You can breed foxhounds from wolves in 200 generations, but you can also breed a solution to a problem. If we don't know the answer to a problem we may guess it. Then we perform the old game of mutation and selection:

• start with a population (~30) of guesses G_k, call them genes
• compute pragmatic information S_k for each gene
• select the best genes (~ 8) with respect to S_k

• mutate ($G_k + \delta G$) and propagate back to original size of population.

The crucial step is selection - a highly nonlinear process -, not random mutation (see Becker;[5] actually evolution can work with deterministic mutations, as well). So even if there is some connection between information and entropy (entropy being the information which we don't need), evolution and entropy production are totally unconnected processes.

The main problems when dealing with such schemes are somewhat hidden. For efficiency it is necessary to find an adequate representation of the system (What is essential? What is a natural mathematical description?) and to choose the appropriate form and size of mutations.

CAN LIFE EMERGE ON A PLANET WITHOUT CARBON?

Carbon owns some remarkable properties which it does not share with any other element.[6] In carbon chemistry, we find:
• many different compounds with irregular but long-range correlations;
• cyclic compounds;
• various types of bindings (C-C, C=C, aromatic, etc.);
• similar binding energies with its neighbours in the periodic table → easy to change partners;
• convenient reaction temperatures;
• existence of cyclic (autocatalytic) reactions.

From this we can guess that self-organisation is not trivial. It may crucially depend on the nature of the building blocks.

CONSTRUCTION, CREATIVITY, AND EMERGENCE

Let me introduce some definitions:
• *construction*: the assembly of given or found building blocks according to given or found construction rules;
• *creativity*: the seemingly effortless construction of an abundance of constructs, which make a sense in a given environment;
• *emergence*: the appearance of new structures, functions, properties etc. (of the impossible!) connected with a (considerable) increase in pragmatic information.

We find a lot of creativity in evolutionary games. We also see (in our simulations) that without a vision there is no creativity. (This is not the place to discuss this issue in natural evolution ... Was Teilhard de Chardin[7] right after all when he said that the goal of evolution was the point Ω?)

Can we hope to find emergence in evolutionary games? Evolution may be able to generate computer programs for specific tasks. However, an algorithm is based on insight, a kind of abstraction. Can we arrive at abstraction as an emergent feature? This cannot be answered at the moment.

COMPLEMENTARY PICTURES OF TIME AND LOGICS

As we have seen from our version of Antigone, logics and creativity are not compatible. Complementarity arises between being and becoming. A. C. v. Müller[8] has argued that there are two complementary pictures of time and logics, and that it is not possible to mix the two pictures because each of them is self-consistent (see Table 1).

Table 1. Two complementary pictures of time and logics.

CLASSICAL	COMPLEX
Linear-successive time (Kant)	Complex time (Bergson)
Logics	Dialectics
Subject-object dichotomy	Ipsoflexivity
Causality	Creativity, correlations, emergence
$dS/dt = 0$	$dS/dt > 0$
Being	Becoming
Security	Happiness

Prediction is only possible in the classical paradigm, i.e. in processes where information is not changed. This coincides with the role of time in physics. In ordinary classical physics and quantum physics, time is treated (not as an observable but) as an external parameter, given a priori. However, in those branches of physics in which information is not conserved (general relativity, entropy production) we work in the complex paradigm: time is generated by processes.

In evolutionary processes, the notion of control gets a new flavour. It can no longer mean forcing the system to stay in some mode of operation. The new notion is connected to the question under which circumstances evolution is optimal, i.e. under which conditions dS/dt is large on a medium to large time scale.

CREATIVITY AND PRECISION

Control in the classical sense obstructs creativity. Luciano de Crescenzo[9] asked the question whether you can have love and freedom at the same time. Unfortunately he did not give a recipe because his question seems to be similar to the one whether you can have creativity and precision in the same system.

In a former paper[10] we suggested a league system. Evolution occurs on each level with a level dependent mutation rate in order to protect the "artists" against the "accountants" by keeping them apart. Following Caianiello[11] we use a homogeneous hierarchical modular selection scheme:

- hierarchical levels 0 ... L ~ "leagues"; L ~ 6 ... 30 ~ "large";
- modular mutation rate $\varepsilon_k = \varepsilon_0 R^{-k}$; R ~ 1.1 ... 4 ~ "small";
- selection and mutation seperately at each level;
- promotion of the best individuals of each level to the next one.

We conclude that you can have love and freedom at the same time - but not on the same level.

CONSERVATIVISM AND INNOVATION

The hierarchical selection scheme has brought considerable progress, in particular in connection with evolutionary crossover. However it has not solved all problems. Since "old" ideas on a given hierarchical level had the chance of becoming optimised during several generations "new" ideas coming up from the lower levels have little chance of being accepted. A Bavarian proverb characterises very well this conservative habit:

's gibt hoid nix bessas ois wia eppas guats (there's nothing better than something good).

On the other hand, if we introduce permanent revolution by replacing the upper-level individuals by the upcomers, we find a permanent replacement of old ideas by new ones - but no idea reaches its maturity.

The "political center" must be somewhere in between. The situation reminds of the critical temperature in phase transitions which is an unstable fixed point of the renormalisation group. It is an open question how to arrive there in an evolutionary game. It may well be that we need an algorithm which is non-local in time.

CONCLUSION

Evolutionary games are well suited to approach complex problems. Their development is unpredictable, or, to be more precise, prediction is limited to a time horizon τ given by the inverse of the learning speed dS/dt. A high value of the learning speed is desirable. This is connected with dialectic synthesis, creativity, emergence, and innovation.

Control in the classical sense is countreproductive. We have to give a new meaning to this term: *the purpose of control is to design rules (of representation, of selection, of mutation etc.) which encourage creativity and innovation in the system without prohibiting maturation of ideas.*

Some rules have been established for this purpose, such as the hierarchical selection scheme; but there are also open questions, such as how to find the political center between conservativism and permanent revolution, or how to facilitate the emergence of higher degree complexity.

REFERENCES

1. Umberto Eco, *Il nome della rosa*, Fabbri-Bompiani (1980).
2. Sophocles, *Antigone*, http://classics.mit.edu/Sophocles/antigone.html (442 BC).
3. Jörg D. Becker, Learning as gain of information, in: G.J. Dalenoort ed., *The paradigm of self-organization II*, Gordon & Breach (1994).
4. Ioannis Antoniou and Jörg D. Becker, *Pragmatic Information and the Misra Entropy Production Operator*, (1999), in preparation.
5. Jörg D. Becker, Entropy and Evolution, in: H. Atmanspacher and G. J. Dalenoort eds., *Inside versus outside*, Springer (1994).
6. Astrid Müller and Lutz Müller, Could life emerge in a world without carbon? *Contribution to the workshop SOP*, Oct 10, Neubiberg, (1998), unpublished.
7. Pierre Teilhard de Chardin, *L'avenir de l'homme*, Edition du Seuil (1959).
8. Albrecht A. C. v. Müller, *Zeit und Logik*, Baur (1983).
9. Luciano de Crescenzo, *Storia della filosofia greca I: I presocratici*, Mondadori (1983).
10. Jörg D. Becker, Creativity and Precision in Evolutionary Synthesis and the Hierarchical Selection Scheme, to appear in *Chaos, Solitons, and Fractals*, (1999), in press.
11. Eduardo R. Caianiello, Some remarks on organization and structure, *Biol. Cybernetics*, 26, pp.151-158 (1977).

PANEL SUMMARY: **CAN LEARNING GUIDE EVOLUTION?**

Panellists: Roberto Bordogna (*chairperson*)[1] and Richard Walker[2]

[1]Media Science & Technology ESAS
IUSS - Pavia's University
[2]Trainet S.p.a.
Rome, Italy

FOREWORD (*R. Bordogna*)

Since ancient time learning has been investigated as a human fundamental capability. "Considering what have been seen" is a specific "human being" function, according to Plato (Cratilo 399 B),[1] for instance. It is often stressed that evolution (ontogenesis and phylogenesis) is the result of heredity and of acquired characters of unity and members of a multiplicity. It is as well usually accepted that "living systems" are both modified and modify their ambits accordingly to contingent necessity and to a given degree of freedom.[2] If we describe this as a learning process, certainly some artificial systems too appear capable of a certain degree of learning.

MAPPING DIFFERENT APPROACHES IN A COMMON FRAMEWORK

It is well known that "learning" and "evolution" may assume very different meanings accordingly to various approaches and contests in which these words are used. For some scholars psychology is a branch of natural science and evolution is just the result of heredity and of acquired characters.[3] Nether the less there are evidences that human beings often express their goals and creativity accordingly to cultural background and that modify their living stile and ambits accordingly to personal belief and knowledge with a degree of freedom which seems to improve with work productivity and social wealth. Human beings' social economic history offers evidence that an increase of productivity and social wealth may support human development and more freedom.[4]

So called knowledge-based economies are considered increasingly based on "human capital" improvement[5] but appear to be related as well to technical means[6] and technologies (as information technology and "media")[7] that support operations, cooperation, learning and various people's interactions in natural, artificial and social commons.[8,9]

In this case not only it seems impossible to apply a disciplinary matrix[10] based on the body-mind "Cartesian cut" approach to the learning process, but mankind development

Human and Machine Perception 2: Emergence, Attention, and Creativity,
edited by Cantoni *et al.* Kluwer Academic / Plenum Publishers, New York, 1999.

appears the result of a very complex civilisation process. This process appears to be based both on individual and group or community basic skills improvement, where learning and creativity take place in increasingly artificial ambits.

These accelerating phenomena appear to include systems and artefacts capable to handle "in process" multimedia data and information, taking decisions and acting thanks (for instance) to an inference-"engine".

It is then important to provide a common framework to the discussion suitable to gather (provisionally at least) heterogeneous scientific, engineering and humanistic research contributions.

The conceptual framework proposed here is based on two couple of very well known and basic antinomies: *multiplicity - unity* and *necessity - freedom.*

Crossing these two couple of concepts on a graph provide one (of the possible) taxonomies that appear to be of some help to describe agent-and-its-operative-ambit relationship that seems common to several disciplines (see Fig. 1), ranging from biology, engineering, to economics and other social science.

In fact deterministic approaches are often interested in general results, that can be applied to a multiplicity (as a particular class of living systems, of groups or of systems and so on), that have character of a necessity (a scientific law) in the given circumstances, while subjectivist approaches tends to emphasise the freedom of the individual (a unity as an "actor" or an "agent") against its operative ambit.

For example, in a software engineering domain, the inheritance mechanism of attribute and functions from a class (a multiplicity) for a given instance (a unity) has character of necessity. More sophisticated systems can make choices from given options, or cope with a particular state of an operation ambit, accordingly to some strategy and to the degree of freedom of the agent.

"Autopoiesis"[11] concept, that describe a "living" systems as a "closed feedback" network of processes, made of components that produce components, "structurally coupled" with the "environment", is accepted by some scholar also in organisational studies domain, but it is a (declared) mechanistic model.

While it is not possible to discuss here in depth the problem of agency in economics it is easy to see that deterministic approaches appear not suitable to model human development. It is known that Adam Smith concept of capitalist stresses its rational strategy in front of the market, and wage as a mean to reinforce labour behaviour, but more modern contributions analyse production ambits in "district" where productivity may leverage out of more "soft" factor, a shared "atmosphere" (of knowledge, skills, technologies, practices and culture).[12] And in fact while "bounded rationality" behavior,[13] due to agent's knowledge limitation, is considered more appropriate to describe behaviours in organisations, more recent contributions stress the role of increasingly differentiated[14]

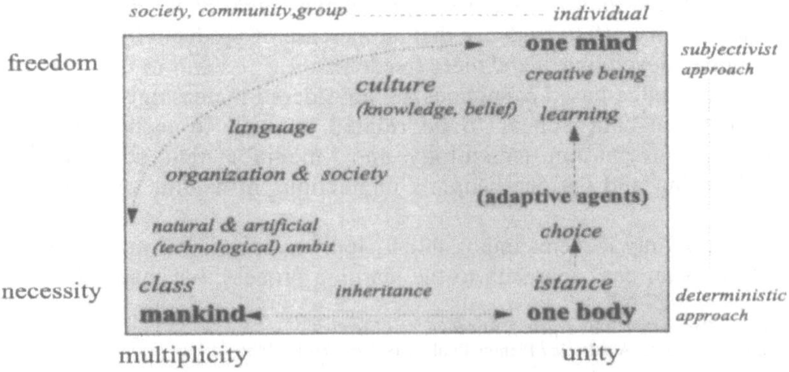

Figure 1. An agent's learning conceptual framework.

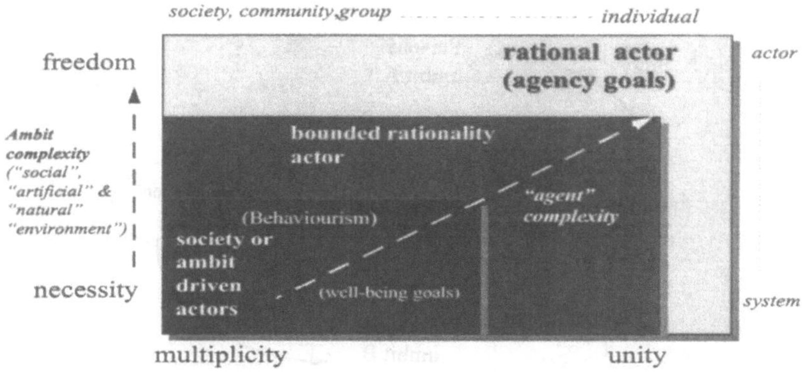

Figure 2. Actor, group conceptualisations.

actors' goals (which act not only accordingly to "well-being" goals but with a more general "agency" agenda)[15] in "social system",[16-19] where cooperation is based on mutual influence,[20] culture[21] and trust.[22] This emphasise the opportunity of a very radical micro-macro approach (ranging from biology to social system) to "learning" processes (driven also "by doing"[17] or "by using"[23]) that is considered here particularly useful to study fundamental social ambits,[24,25] such as markets.

In fact markets may be considered learning ambits where agents interact accordingly to social practices and institution (such as bidding style, currencies and other influences - such law and power) with social commons (as systems), using a standard communication ambit to define a common judgement: prices (see Fig. 2).

AMBIT AND COMMON AMBIT ARE RELATIONAL

It is sufficient to analyse a generic communication ambit to see that the (provisionally) proposed taxonomy must be identified on a "by process" / "by context" base. In fact unity and multiplicity must be considered a contingency because a person may be for example member of several communities at any given time ("is-a" father, a postmen and an Italian for instance) and because also actor's degrees of necessity and freedom are contingency defined.

When one makes a phone call the forward communication "channel", made with a given technology that empowers the addressed person's biology and culture, has a character of necessity for a person that must use it, in order to hear the message. Is an ambit in which information as an entity[26] may be exchanged when there is a common communication context between the agents.

But the actor may decide to express its freedom to leverage on his (at the time given) biology, knowledge and culture on a different way, for instance writing back a letter, so using a totally different social and technological system to complete the communication transaction. This last has character of necessity for the original issuer of the first message. This dynamic can be described as a "trial and error" process to experiment "reality". Because of what we have said the empiric experience appear to be based on (given) social assumptions and (contingent) constraints as well as on the freedom of the actor agenda. This contingent subject - object switch, in space and time (where the agent expresses its freedom or vice-versa its behaviour may be related to some type of necessity), empowered by artefacts, seems to be a basic social relationship[8,9] and a foundation of several learning processes. Because of all that it may be useful to define a generic (anthropic) "continuum" ambit called *agent* (or personal) *ambit* and *common ambit* its shared part.[8]

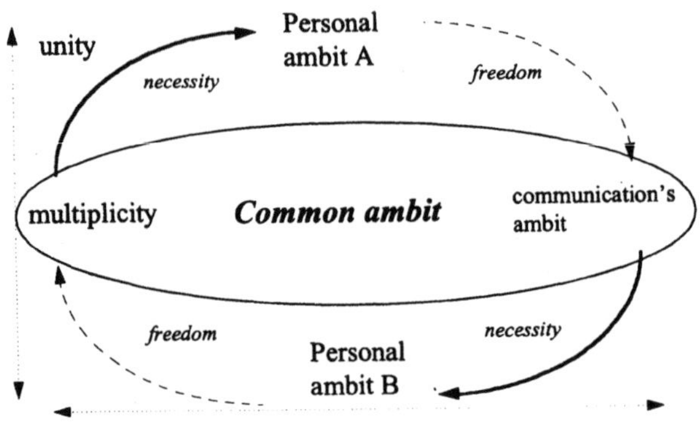

Figure 3. Social-economic development and learning process in Personal/Common ambit must be seen as a continuum.

Agent (personal) ambit is an attempt to escape from any predetermined taxonomy and "ontology" and does not make any assumption on the nature of the (living) system.

Because common ambit is defined as a shared agent ambit, it can be identified only with reference to another agent (personal) ambit (or to a class of ambit that shares that common ambit - see Fig. 3).

Personal ambit can be modelled as a contingently suitable vector space (of natural, artificial and cultural/social variables or functions).

A GENERAL THEORETICAL FRAMEWORK – "STEPPING STONES"
(*R. Walker*)

It is always useful to begin by defining terms. So, in what follows, I will take the term "evolution" to mean modifications of structure or behaviour via the action of natural selection on *populations* of individuals with differing levels of reproductive fitness. And when I refer to "learning" I will mean adaptations during the life-time of these *individuals,* optimising some kind of second order "objective function" - itself a product of evolution. Defined in this way it is immediately obvious that learning is by no means a human prerogative and that in reality a broad range of "Complex Adaptive Systems" exhibit learning. In what follows I will concentrate on two examples: biological organisms evolving and learning within a natural environment – and firms, evolving and learning within a capitalist economy.

To understand the way in which biological organisms or firms evolve and learn – the way in which learning helps evolution and evolution directs learning - it is useful to look at I'm going to call the four "Cs": Change, Choice, Complexity and – this is a last minute addition after hearing the presentations this morning – Creativity.

We need to look at change because real-world environments are in a state of continual flux. Biological organisms have to deal with changes in temperature and in rainfall, the introduction of new predators, the extinction of prey species. Firms have to deal with changing prices for inputs – labour and energy for instance – changing patterns of demand, technological innovation and obsolescence, the birth of new competitors and the demise of old ones. No organism and no firm ever lives twice in the same environment.

Choice is equally important. Just as in Eastern philosophy there are many roads to wisdom, so there is more than one way in which organisms can achieve viability: biological organisms like Darwin's finches can occupy different niches; firms can adopt different market strategies. In short biological organisms and firms do not just *react* to an environment in some pre-determined way. They make *choices.* And like change these choices play a fundamental role in the evolutionary process. It is *change* and *choice* – I will argue -which make complexity and creativity possible – even if only in certain very specific conditions – which we will come to later.

The general argument will proceed along two parallel tracks: on the one hand I will briefly illustrate a general theoretical framework which can help us to understand what is going on. On the other I will be presenting computer simulations of evolution and learning– some my own – some from other workers in the field, interpreting the results in terms of the theory. Finally I will combine simulation and theory to develop a number of falsifiable, counter-intuitive conjectures about the functioning of evolution and learning in specific kinds of Complex Adaptive System.

In many theoretical discussions evolution is depicted as taking place in a static environment. This environment defines a so-called "fitness landscape"[27] associating every possible genotype with a particular "fitness value" (the z-axis on the graph). In this vision evolution consists of "hill-climbing". At every stage in evolution populations move from regions of the landscape with low fitness values to neighbouring regions with higher values: biological organisms produce higher numbers of viable offspring, companies become more efficient and more profitable.

The hill-climbing view succeeds in explaining the optimising function of evolution but fails to take account of environmental change. When the climate changes previously well-adapted organisms can go extinct; when new competitors arrive or markets or technologies change successful firms can fail. In other words environmental change can often be damaging. This means that real world organisms and firms do not just climb hills; it would be more realistic to depict them as riding on a roller coaster where *reductions* in fitness can be just as common as *improvements.*

In order to visualise the role of environmental change it is helpful to replace the classic fitness landscape with an *extended fitness landscape*[28] showing the fitness of different genotypes not just in *one* single environment but in *any conceivable* environment. In such an extended fitness landscape each point represents a specific genotype inhabiting a specific environment.

Extended fitness landscapes (like classical ones) occupy a high dimensional space which it is not possible to represent graphically. For heuristic purposes it is useful therefore to project the landscape onto a simple three dimensional space – where the x-axis represents the genotype, the y-axis represents the environment and the z-axis stands for the fitness of the combination.

The vital requirement for evolutionary lineages is not that they should *optimise* their fitness but that they should always be *viable.* To highlight this aspect of the evolutionary

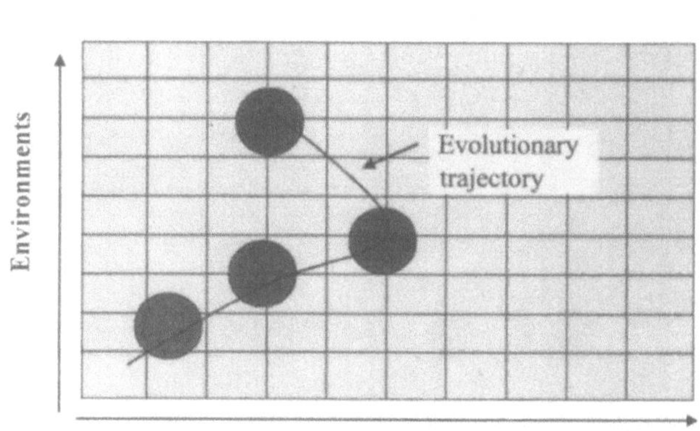

Figure 4. The stepping stone model (basic framework).

process we can take a horizontal slice through the landscape, drawing it as a *lake* with *stepping stones* - where the stepping stones represent viable combinations of genotype and environment and the water represents non-viable combinations (see Fig. 4). In this picture evolution consists of a trajectory from stepping stone to stepping stone The distance between the stones can be interpreted as the "cost" of moving from one stepping stone to the next, i.e. the difficulty of modifying the genotype of a biological organism or of changing the basic organisational structure of a company; the cost of moving from one environmental niche to another or of abandoning a market and moving into another one. If the distance between two stepping stones is too great it is impossible to cross the lake.

SIMULATING EVOLUTION AND LEARNING AND INTREPRETING THE RESULTS

One of the most effective ways to investigate the relationship between learning and evolution is computer simulation. Since the late 1980s a number of workers, including myself, have simulated the evolution of populations of artificial organisms, modelled as "neural networks". Each organism is assigned a "task". Its ability to perform this task defines its *fitness:* the organisms which perform the task best produce a higher number of offspring than less effective organisms. Reproduction introduces variation into the population in the form of "mutations" affecting the neural architecture and/or the value of synaptic connections. This cycle of learning/selection/reproduction is repeated many times simulating the *evolution* of an optimal architecture via natural selection. Optionally organisms can be "trained" using classical neural network learning algorithms. This is *learning*. By comparing the respective performances of simulations with and without training it is possible to gain fresh insight into the relationship between learning and evolution.

Simulation results show that if artificial organisms are assigned extremely simple tasks (e.g computing Boolean functions in two variables)[28] and allowed to evolve in a stable environment learning is unnecessary. Even with the learning algorithm switched off the Darwinian mechanism incorporated in the model is powerful enough to create architectures capable of achieving a 100% success rate.[28] This result is easy to interpret in terms of the extended fitness landscape. The optimal neural architecture for implementing elementary functions is simple and can be easily built up in incremental steps; the environment is by definition a static one. The evolutionary trajectory (see Fig. 5) can thus be visualised as a sequence of small (easy) jumps between stepping stones located along the horizontal axis.

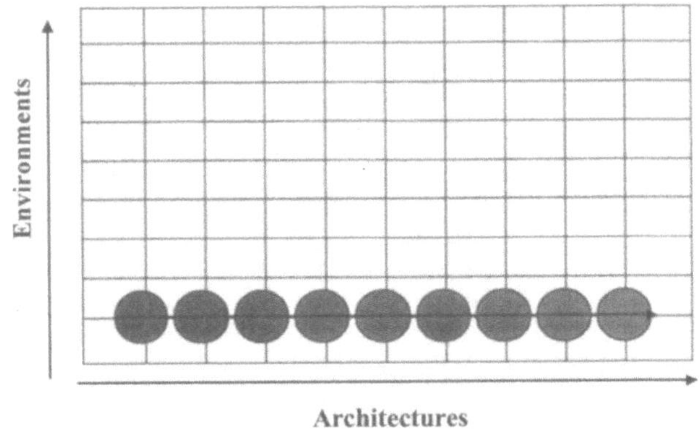

Figure 5. Linear evolution in a stable environment.

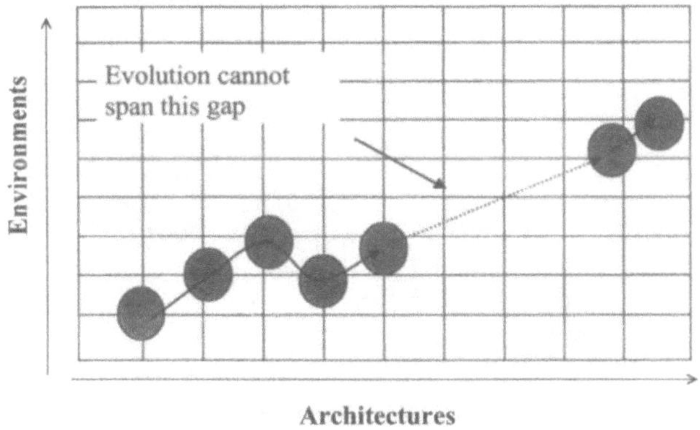

Figure 6. Evolution with fast environmental change.

In a series of classic experiments Hinton and Nowland[29] have created simulations of evolution and learning where the tasks to be acquired were more complex than those used in the experiments just reported. Nolfi and Parisi have built on this work[30] extending the basic model to include a changing environment. In both cases the levels of fitness achieved were higher when organisms were allowed to "learn" as well as to "evolve". These simulations emulate the so-called "Baldwin effect"[31] in which learning (or other forms of developmental plasticity) "guides" evolution.

The contribution of learning to evolution can again be interpreted in terms of the stepping stones. In pure Darwinian evolution, without learning, a given neural architecture is viable only for a limited range of environments. As a result rapid environmental change can create gaps between the stepping stones which evolution finds it difficult or impossible to jump (see Fig. 6). With learning, on the other hand, the same initial architecture can achieve high levels of fitness in many different environments – we stretch the stepping stone in a vertical direction; similarly learning can allow organisms with different initial architectures to achieve viability in a given environment – this stretches the stepping stone horizontally (see Fig. 7). The overall effect is to make the stepping stones bigger, to decrease the distance between stepping stones and thus, as we see in Figure 8, to reduce the probability of getting stuck in an evolutionary *cul de sac*.

THE ROLE OF "CHOICE" – NICHE SELECTION

Despite the success of many research groups in simulating the Baldwin effect in the acquisition of simple learning tasks the combination of evolution and learning is far from

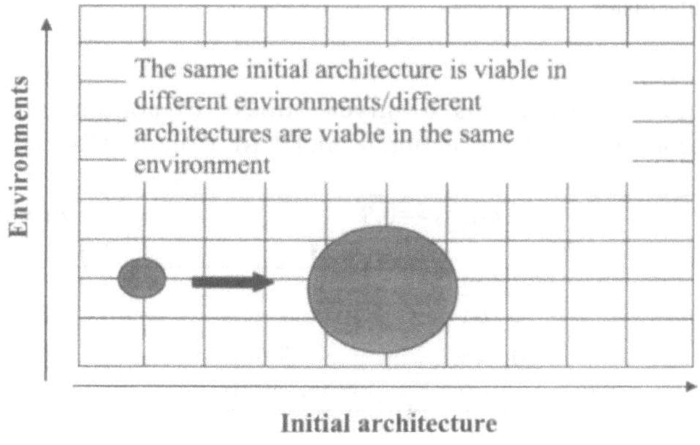

Figure 7. Increasing the size of the stepping stones.

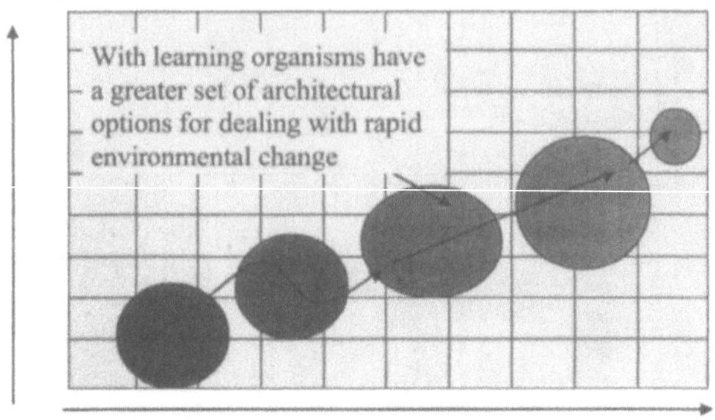

Figure 8. Combined evolution and learning with fast evolutionary change.

being a panacea. In some of my own simulation work[28] the combination of evolution and learning unexpectedly failed to produce artificial organisms capable of computing apparently simple Boolean functions in three variables. Given the evident success of biological organisms in carrying out far more complex tasks it was obvious that the simulation model failed to match biological reality. Something was missing.

Waddington[32] and other later authors[33] have suggested that in reality Darwinian natural selection is only one side of the grand evolutionary picture. In practice it is not only the environment which "selects" fit organisms. At any one time biological organisms belonging to a species will inhabit a number of different environments, achieving higher levels of reproductive fitness in some than in others. The overall population thus "selects" its preferred niches. This "niche selection" complements "natural selection". It seems intuitively likely that this will favour the evolutionary process: if a population of organisms is allowed to select its own task (the task best fitted to a particular genotype or organisational structure) it is likely to adapt far more easily than if the task is randomly chosen by exogenous environmental forces.

This hypothesis is again susceptible to investigation via computer simulation. In one model[34] each organism in the population was assigned a different initial task which was "inherited" by its offspring. Tasks were subject to selective pressures and to random "mutation" in the same way as neural architectures: where organisms were well adapted to their tasks, the task spread through the population – poor adaptation, on the other hand, lead to elimination.

Simulations results show that for tasks of a given complexity the "niche selection" model produces "fitter" organisms than a control model based on a more conventional combination of Darwinian selection and learning (see Table 1). In certain conditions (to be examined later) the niche selection model produces organisms capable of acquiring relatively complex tasks.

Table 1. Number of organisms of given fitness in populations of homogeneous complexity. Legend: NS: Niche Selection Model, CM: Control Model

	Fitness>=0.9		*Fitness>=0.8*	
Complexity	NS	CM	NS	CM
0-4	107,422	503	137,202	1,733
5-8	5,493	1,036	8,484	2,256
9-12	10,245	3,666	19,772	7,802
>12	0	0	4	48

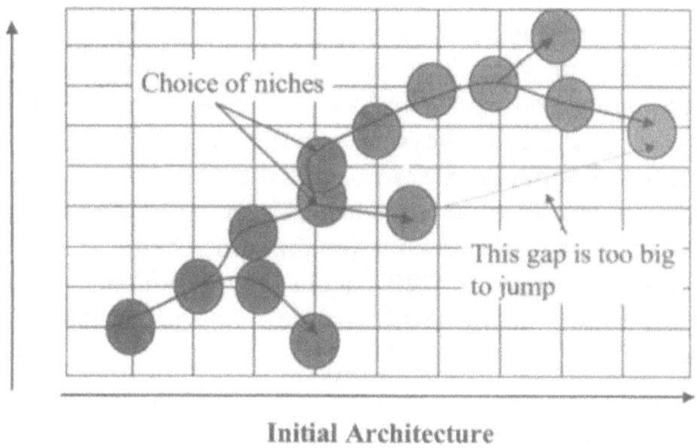

Figure 9. Niche selection.

The results of these simulations using niche selection can again be interpreted in terms of stepping stones. Without "niche selection" environmental change constitutes an external driving force, obliging organisms to adapt. If they are not capable of doing so – if the gap between the stepping stones is too wide – they perish. Niche selection, on the other hand, allows bifurcations. A part of the population can follow one evolutionary trajectory – another part can follow another (see Fig. 9). The probability that this process will produce at least one successful lineage is far higher than it would have been in the case of forced adaptation to a single set of environmental requirements.

COMPLEXITY, CREATIVITY AND ENVIRONMENTAL CHANGE

The original purpose of introducing "niche selection" into the simulations was to investigate whether this could facilitate the acquisition of tasks too complex for a conventional combination of Darwinian evolution and learning - to see whether a model incorporating niche selection could demonstrate some hint of the creative, complexity-building power of natural evolution and learning.

Further results from the Niche Selection Model demonstrate that this is possible – in the right conditions. Analysis of the experimental data shows that the evolution of complexity depends critically on the *rate of environmental change* (in the simulations the probability with which tasks were modified between one generation and the next). With very slow rates of change the whole population came to be dominated by organisms

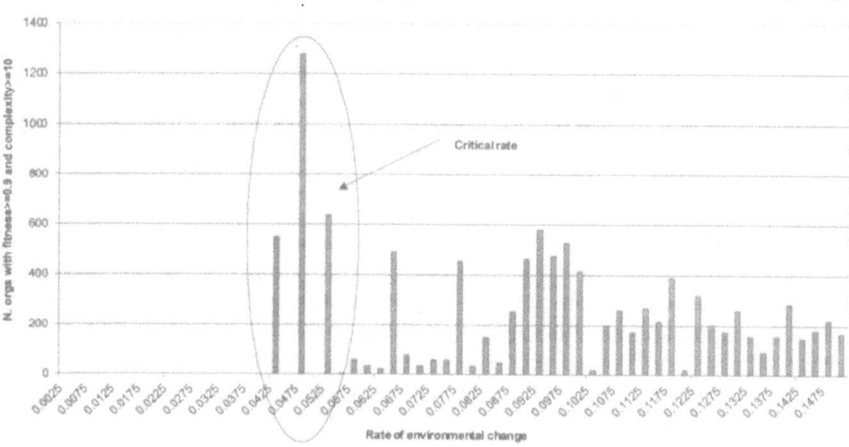

Figure 10. Number of fit, complex organisms for different rates of environmental change.

performing extremely simple tasks. At high rates of change, on the other hand, the population failed to keep up with the continual change in tasks and fitness levels fell to levels which in a biological setting would have spelled extinction. The simulations showed however that rates of change just beyond a mathematically well-defined critical value produced a differentiated population of *fit organisms*. Although the majority continued to perform simple tasks, a small proportion successfully performed *relatively complex* ones (see Fig. 10). At least in the model change (or rather a specific rate of change) and choice (in the form of niche selection) were necessary conditions for the emergence of complexity. With too little – or too much change – or without choice, evolution failed to play its expected creative role.

CONJECTURES AND CONCLUSIONS

The results obtained from the simulations suggest the conjecture that creativity may depend on a critical rate of change, not just in the artificial setting of the model but also in real life complex systems: biological organisms or firms or ecosystems. If this conjecture is correct a static environment – the sort of environment you could find on the bottom of the sea bed, in a feudal economy, in over protected forests in national parks – will produce evolutionary simplification and homogenisation. Complexity will only emerge when the environment is disturbed in some way. Looking backwards in time it is not difficult to find possible instances of this effect at work. In the case of biological organisms there is considerable evidence that major evolutionary change is connected with serious disturbance of the environment by impacts with extraterrestrial bodies (e.g. the impact alleged to have ended the Cretaceous period – and the dinosaurs). In economic history it is easy to find instances of major economic change following demographic upheaval (e.g. the Black Death in 14th Century Europe) or major price shocks (e.g. the explosive rise in oil prices between 1973 and 1980). In ecology it is commonly argued that disturbances to the environment (e.g. forest fires) play a vital role in maintaining diversity.

In order to verify the conjecture it is nonetheless necessary to proceed quantitatively – and this is entirely feasible. It would be possible for instance to produce time series in which the instability of key input or output prices (the rate of environmental change) is plotted against an index of economic diversification (complexity) and economic growth (fitness). If the conjecture is correct, periods of stable prices should be associated with falling diversity and high growth. Rates of instability just beyond some critical threshold should, on the other hand, be associated with increasing diversification (albeit with somewhat lower growth). Even higher rates of instability, it is predicted, will be associated with high levels of diversification but very slow (or negative) growth. Similar tests are possible for ecosystems, where complexity could be measured by species numbers and the rate of environmental change by climate instability (for studies based on historical or paleontological records) or by the number of forest fires or floods or other catastrophes (for shorter term investigations). Here again it is predicted that low levels of environmental change will lead to simplification, higher rates to complexification, still higher rates to destructive loss.

To sum up: it is conjectured that evolutionary creativity depends on complex systems' freedom to choose their own strategy and on the rate of environmental change. Complexity, it is suggested, can only emerge when this rate is just beyond a well-defined "critical value". The conjectured relationship between change, choice and complexity could easily be disproved by empirical investigation; the conjecture therefore constitutes a valid scientific hypothesis in Popper's sense of the term. The empirical investigation of the hypothesis opens an interesting field for future research.

DISCUSSION

G. Rigamonti question to R. Bordogna: You said that there is a positive correlation between economic development and human development. What do you mean precisely with this word?

R. Bordogna: I use this concept as stated in the United Nations report on Human Development.[4]

G. Rigamonti question to R. Walker: Do you think that evolutionary paths could contain loops? Can a species return to the same position after a number of steps? Can two different evolutionary paths converge to the same position?

R. Walker: As far as concerns physical (as opposed to logical) organisms and environments the strict answer to your question about loops should be "no". Given that environments and organisms are in a state of perpetual flux it is extremely unlikely that history will ever repeat itself – at least not precisely. If we take a broader view it is obvious however that roughly cyclical trajectories are a biological and economic reality. When there are short term cyclical changes in the environment (e.g. seasonal change, business cycles) individual organisms may adapt to new conditions via learning but then re-adapt when the environment returns to its original state. We would draw this trajectory as a cycle on the extended fitness landscape.

Coming to your second question about convergent evolutionary paths the answer is partially "yes": at the phenotypical level convergent evolution is a real biological phenomenon – very distant evolutionary lineages can produce similar architectural or behavioural solutions to similar environmental challenges – think for example of the development of wings in bats and birds. At a genotypical level on the other hand every evolutionary tree is unique. Even for relatively primitive organism the genome is so complex that chance of the branch of one tree "touching" the branch of another is vanishingly small.

V. Cantoni: In your simulations what was the number of organisms produced for different complexities? Is it possible that the number of organisms could be inversely proportional to the square of complexity?

R. Walker: I don't know the answer to your question but it is an interesting hypothesis which it should be easy to check with the experimental data.

Note written after the meeting: Counting the number of fit organisms in different complexity classes at rates of environmental change just beyond the critical threshold, we obtain the results seen in Figure 11. The data appears to be compatible with Virginio Cantoni's hypothesis although there are too few data points for rigorous confirmation.

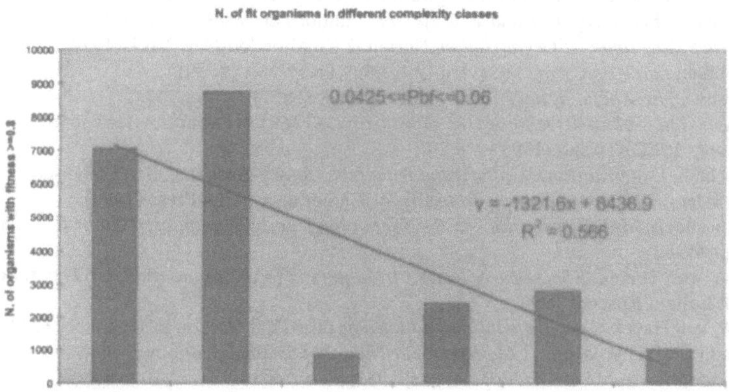

Figure 11. N. of fit organisms in different complexity classes.

POSSIBLE GENERAL CONCLUSIONS *(R.Bordogna)*

As general conclusion it seems possible to say that the degree of freedom and learning increases with the complexity of both the "agent" and his "ambit" (both natural, artificial and cultural).

During the discussion it have been shown that the relational "continuum" agent-and-his-ambit (natural, artificial, cultural /social) that have been called agent (personal) ambit allows to identify a common ambit with reference to another agent ambit (or to a class of personal ambits) on a contingency base, avoiding any predetermined definition and subject-object relationship, can be modelled as a vector space suitable for that contingency.

In the proposed paradigm social development is associated to efficiency and efficacy of an incremental circular process between personal and common ambit that could be improved by multimedia information and communications. Technologies and education are condition necessary but not sufficient for development because a fundamental development factor appears to be personal ambit strategies: social values and entrepreneurship driven (or not), with which a person expresses his freedom in a global economy.

REFERENCES

1. Platone, *Tutti gli scritti,* Giovanni Reale ed., Rusconi, Milan (1996).
2. Richard L. Gregory, *The Oxford Companion to the Mind,* Oxford University Press (1987).
3. V Cantoni, V. Di Gesù, A. Setti, D. Tegolo eds., Human & machine perception - Information fusion - Plenum Press N.Y. (1997).
4. United Nations Development Programme, UNDP *Human development report,* Oxford University Press 1996 - It. Ed. Rosenberg & Sellier (1996).
5. OECD, - Angus Maddison, *Monitoring the World Economy 1820-1992*, Paris (1995).
6. A.Donald Norman, *Things that make us smart*, Addison-Wesley USA (1993).
7. OECD, *Information Technology Outlook,* Paris (1997).
8. R. Bordogna, *Improving personal & common ambit for "reality" experimenting/learnig-producing/creating to accelerate social and economic development.* 2nd Alps, Bologna (Italy) Euroconference - European Union, ENIMM UK (1996).
9. R. Bordogna, *Leveraging on multimedia technologies to modernize social economic ambit development paths*, In automated systems based on human skill, D.Brandt ed (IMA/HDZ-RWTH Aachen - Germany & Stefan Institute Ljubljana Slovenia), 6^IFAC Symposium, Kranjska gora, Slovenia (1997).
10. Thomas S. Khun, *The structure of scientific revolution*, University of ChicagoPress, Chicago (1970).
11. H.R. Maturana and F.J. Varela, *Autopoiesis and Cognition. The realization of the Living,* Reidel Dordrecht-Holland. (1980).
12. Alfred Marshall, *Principles of economics* (1890), Mac Millan University Press, Cambridge (1961).
13. Herbert Simon, *The Sciences of the artificial*, MIT Press, Cambridge (Mass.) (1981).
14. Norbert Elias, *Die Gesellschaft der Individuen*, Suhrkamp, Frankfurt (1987).
15. K. Amartya Sen, *Inequality Reexamined,* Oxford University Press, Oxford (1992).
16. M. Crozier and E. Friedberg, *L'acteur et le système,* Editions du Seuil, Paris (1977).
17. Kenneth Arrow, *The limits of Organization*, Norton & Co., London-New York (1974).
18. Alfred Chandler, *Scale and Scope*, Harvard University Press, MA (1990).
19. Talcott Parsons, *The Social System* The Free Press, USA (1951).
20. Weick Karl,. *The Social Psychology of Organizing,* (1969) Newbery Award Records, Italian edition *Organizzare* - ISEDI Torino (1993).
21. Edgar H. Schein, *Organizational Culture and leadership,*Jossey-Bass Inc., San Francisco (1985).
22. Anthony Giddens, *The consequences of Modernity*, Cambridge, Polity Press (1990).
23. Nathan Rosenberg, *Inside the black box: Technology and Economics,* Cambridge University Press, Cambridge (1982).
24. Friedrich A. von Hayek, *The sensory order. An inquiry into the foundations of Theoretical Psychology* (1952), It. Edition Rusconi 1990.
25. Friedrich A. von Hayek, *Law, Legislation and Liberty,* (1982), It. edition Il Saggiatore (1994).
26. C.E. Shannon and W. Weavers, *A Mathematical Theory of Communications* (1964).
27. S Wright, Evolution in Mendelian Populations, 1931, *Genetics* 16:97.
28. R. Walker and H.H. Lund, "Step by step" evolution of cognitive function: the Composite Learning System Model, submitted to *Biological Cybernetics* (1995).

29. G.E. Hinton and S.J. Nowlan, How learning guides evolution, *Complex Systems, 1, 495-502* (1987).
30. S. Nolfi and D. Parisi, Learning to adapt to changing environments in evolving neural networks, CNR Technical Report 95-15 (1995).
31. J.M. Baldwin, A New Factor in Evolution, *The American Naturalist* 30, 1896, reprinted in "Adaptive Individuals in evolving populations, R.K. Belew & M. Mitchell (eds.), Addison Wesley (1996).
32. C.H. Waddington, *Evolution of an evolutionist,* Ithaca, New York, Cornell University Press (1975).
33. R.K. Belew and M. Mitchell, "Introduction" in *Adaptive Individuals in Evolving Populations* R.K. Belew and M. Mitchell eds., Reading Massachusetts, Addison-Wesley (1996).
34. R. Walker, "Niche Selection" and the Evolution of Complex Behavior in a Changing Environment- a Simulation, submitted to *Artificial Life* (1998).

THE EMERGING PROPERTIES OF NEURONAL NETWORKS: FOCUS ON THE CEREBELLUM

Egidio D'Angelo

Istituto di Fisiologia Generale
Università di Pavia
Via Forlanini 6, I-27100, Pavia, Italy
and
Istituto di Fisica della Materia (INFM), Pavia Unit
Via Forlanini 6, I-27100 Pavia, Italy

INTRODUCTION

A central issue in Neuroscience is how elementary properties of neurones and synapses are related with neuronal network computation, and ultimately with cognition and behaviour.[1-3] In this chapter, I will consider neurones and synapses of the cerebellum, a brain structure of primary importance for co-ordinating movement.[4-6] Figure 1 shows the basic neuronal organisation of the cerebellum, and its connections with some extracerebellar structures. Attention will be focused on the synapse between mossy fibres and granule cells (mf-GrC relay), since recent findings suggest that it may play a more important role for cerebellar computation than previously thought.

THE CEREBELLUM: A COMPUTATIONAL VIEWPOINT

Since the first description of its highly organised neuronal architecture at the beginning of this century, the cerebellum has attracted the interest of scientists investigating the principles of neuronal computation.[4-6] The cerebellar function has initially been inferred from neurological observations and lesion experiments.[7,8] Cerebellar damage causes ataxia, a syndrome characterised by ipotonia, dismetria, tremor, and motor incoordination. The disturbance of voluntary movement suggests that the cerebellum helps the predictive computation required to decelerate movement close to the endpoint of a trajectory, and to determine a smooth transition from one phase of movement to another. The cerebellum

Human and Machine Perception 2: Emergence, Attention, and Creativity,
edited by Cantoni *et al.* Kluwer Academic / Plenum Publishers, New York, 1999.

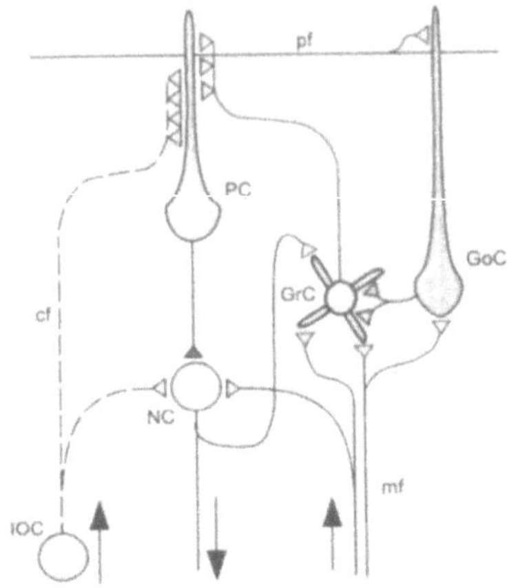

Figure 1. The mossy fibre - granule cell relay of cerebellum. mf, mossy fiber; cf, climbing fiber; pf, parallel fiber; GrC, granule cell; PC, Purkinje cell; NC, nuclear cell; IOC, inferior olivary cell; GoC, Golgi cell; SC, stellate cell; BC, basket cell. Solid lines denote the mf input, dotted lines the cf input. Neurones and synapses in gray are inhibitory. Arrows indicate the flux of information in and out of the cerebellum. Note recurrent GrC activation by NC (after Arbib et al.[3]).

should therefore be able both to determine the proper timing of movement, and to learn appropriate sequences of movements for subsequent retrieval.

Timing and Learning

Based on anatomical and functional considerations[5] two computational models have been proposed, viewing the cerebellum either as a *timing* machine[9,10] or as a *learning* machine.[11,12] Both models, which have polarised neurophysiological thinking about the cerebellar function, were based on the double innervation of PCs by pfs and cfs, although in a quite different way.

The Braitenberg's model[9] is based on the concept of coincidence detection of pf and cf activity by PCs, and does not consider learning. The proper timing of PC response is determined by conduction delays in pfs. However, pfs revealed too fast to act as delay lines. Using Braitenberg's words in a follow-up of his hypothesis, "what is left is the idea that the cerebellar cortex reacts specifically to sequences of input signals when they are presented along a folium or beam in the right tempo".[10]

The Marr-Albus' model[11,12] is based on the assumption that a PC encodes an elemental movement. The PC would work as a perceptron, the double input from pfs and cfs serving to store memory at the pf-PC synapse through a mechanism of associative learning. The cf should carry an error signal acting to change pf efficacy until the error is minimised. Subsequent investigations of the pf-PC synapse confirmed the existence of a form of synaptic plasticity called long-term depression (LTD).[13] However, the physiological implications of LTD remain unclear, and genetic mutations affecting LTD produced contradictory effects on motor learning and memory (see De Shutter[14] for critical review).

Functional Organisation of Cerebellar Inputs

With the aim of establishing a relationship between structure, function, and dynamics of the cerebellar network, several models have been developed (see Arbib et al.[3] for review).

In particular, the Boylls' model emphasised GrC recurrent excitation through cerebellar and precerebellar nuclei,[15,16] while the Blodel's model emphasised the fact that the cf input is organised in sagittal bands orthogonal to pfs providing the basis for *dynamic selection* of patches of PCs revealing coincident cf and pf activity.[17,18] PC patches are activated synchronously, probably exploiting synchronisation of inferior olivary neurones and cfs.[19,20]

It is interesting to note the different type of information conveyed through mfs and cfs in the models. Mfs are generally assumed to convey spatially organised information on the kinematics and metrics of movement through multi-sensorial inputs as well as through a central representation of the movement itself.[6] The role of cfs is less clear. Arbib proposed that cfs specify muscle groups.[15] In Blodel's model,[17] the cf input is sensitive to the context in which movement is generated. According to Llinas,[19] the principal role of cfs is to generate an internal computational state that serves as a context for incoming information through mfs.

SYNAPTIC AND NEURONAL PROPERTIES OF THE MOSSY FIBRE INPUT TO THE CEREBELLUM

Recent advances in cellular neurophysiology are providing a growing body of evidence that the biophysical properties of neurones and synapses can strongly influence network computation.[1-3] Neurones show characteristic discharge patters depending on their own morphology and on the complement of ionic membrane channels, determining the transfer function at any specific synaptic relay.[21-24] Synapses show activity-dependent changes in their efficiency, a property called synaptic plasticity. Two forms of synaptic plasticity, long-term potentiation (LTP) and long-term depression (LTD), are indeed considered the basis of learning and memory in neuronal networks.[24,26]

Surprisingly, although it represents the major input to the cerebellar cortex, very few was known on the mf-GrC relay at the cellular level. In the two following sections I will summarise the classic ideas on the functional properties of mf-GrC relay,[27] and their extension in view of recent experimental observations.

The Marr's Theory of the Granular Layer

Some computational properties of the mf-GrC relay descend directly from its structural organisation.[11,12] The mf-GrC relay can be viewed as a network suitable for spatial pattern separation, which can distinguish activity in separate groups of mossy fibres although they partially overlap. The mf-GrC relay has a peculiar geometric arrangement, in that each mossy fibre diverges on average onto 28 GrCs, and each GrC is contacted by 4 mfs. Functionally, a *codon* is defined as a subset of active mossy fibres needed to fire a granule cell. Spatial pattern separation will be the better the lower the codon size. In simpler words, the probability that a casual combination of active mossy fibres fires a granule cell will be the lower the higher the number of fibres a granule cell has to pick up to be excited. Spatial pattern separation would be improved by lateral inhibition exerted by Golgi cells, and by the sparse representation generated by the patchy distribution of mfs in the cerebellar cortex.[2,3,5,6]

Spatial pattern separation was central to the Marr-Albus theory of the mf-GrC relay. However, three assumptions remained speculative. First, Marr contended that a mf-GrC synapse "whether or not it was initially facilitated, it almost certainly will be at a certain time". And since a uniformly distributed facilitation (equivalent to a decrease in the codon size) would impair spatial pattern separation, he assumed that the mf-GrC synapse was not modifiable. Secondly, the codon size was assumed to be greater then 1, since with a codon size of 1 coincidence detection would be very inefficient. Thirdly, no evidence was provided about the transfer function of granule cells. All three assumptions are worth being experimentally tested, since they are all but irrelevant to understand how the mf-GrC relay participates to cerebellar computation.

New Insight from Experimental Investigations at the Mossy Fibre – Granule Cell Relay

- *Codon size.* We have directly measured the number of mfs needed to fire a granule cell using intracellular recordings.[28] The codon size turned out to be 3 in most granule cells activated from near their resting membrane potential.[11] This result is in keeping with Marr's assumption.
- *Plasticity.* We have recently demonstrated that the mf-GrC relay develops LTP following high-frequency mf activity.[29-31] As at other central synapses, mf-GrC LTP requires activation of NMDA receptors by associative paradigms requiring simultaneous pre- and postsynaptic activation (see next section).[25,26] Critical regulatory factors for LTP induction are the intensity, frequency, and duration of mossy fibre activity, and the regulation of granule cell excitation through the inhibitory Golgi cell circuit (D'Angelo, Armano and Rossi, unpublished observations). The present evidence for mf-GrC LTP is in contrast with the Marr's assumption that the mf-GrC synapse is not modifiable.[11]
- *Input-output relationship.* The granule cell has been so far proposed to correspond to a MacCullogh-Pitt neurone, that is a neurone with a linear input-output relationship.[32] Although this property of GrC excitability can be faithfully reproduced in mathematical models of electroresponsiveness,[32-34] three observations suggest that the electrical GrC behaviour may be more complex than previously thought. First, the GrC can switch to bursting (D'Angelo et al.[33] and Fontana, Naldi, and D'Angelo, unpublished observations), a firing mode which may be important in controlling network synchronisation. Secondly, the inhibitory neurotransmitter GABA can shift the input-output relationship by changing the input conductance of the neurone.[35] Finally, a strong non-linearity in the input-output function of the mf-GrC relay arises from subthreshold integration of impulses, owing to the voltage-dependent behaviour of synaptic channels associated with NMDA receptors.[28]

NMDA Receptors, Mossy Fibre – Granule Cell Transmission, and Motor Behaviour

The main outcome of these experimental investigations is that the mf-GrC relay is plastic and non-linear, differently than previously assumed. These properties largely depend on NMDA receptors and the associated channels. The NMDA channels are voltage-dependent, in that they generate a remarkable conductance only when the postsynaptic membrane is depolarised. Conducting NMDA channels are highly permeable to Ca^{2+} ions, which activate intracellular biochemical cascades regulating plastic and adaptive processes. Depolarisation usually depends on synaptic potentials and spikes revealing the incoming mossy fibre activity. It turns out that the NMDA receptors, by detecting the coincidence of pre- and postsysnaptic activity, implement the Hebbian rule at the molecular level.[25] In addition, NMDA receptors are critical to determine the firing rate of the cell during repetitive transmission.[28] Thus, NMDA receptors are at the core of several processes including non-linear integration of synaptic impulses, regulation of high-frequency transmission, and induction of plasticity (LTP). What is intriguing is that during LTP the NMDA receptors are themselves potentiated, suggesting that they operate at the core of a feed-forward loop, and that synaptic transmission and plasticity are strongly related one to the other.[30]

It is plausible that alterations of NMDA receptors impair motor co-ordination and learning. Although spontaneous mutations have not been revealed so far, NMDA receptors can be mutated using techniques of genetic engineering. NMDA receptors proved valuable tools for a number of reasons: i) at the developmental stage at which LTP has been demonstrated, cerebellar NMDA receptors are prominently expressed in granule cells; ii) NMDA receptors are eteroholigomeric proteins composed of NR1 and NR2 subunits. While the NR2B isoform is expressed at immature stages, the NR2A and NR2C isoforms are expressed at mature stages of development;[36,37] iii) NR2C is exclusively expressed in granule cells.[36,37] Thus, knock-out of the NR2A and NR2C subunits should specifically impair mf-GrC relay functions, with marginal disturbance of cerebellar ontogenesis and of other cerebellar circuits. Indeed, NR2A and NR2C knock-out as well as the deletion of the intracellular part of NR2A and NR2C impaired motor co-ordination and learning in mice.[38–39] It seems therefore that NMDA receptor-mediated synaptic transmission and plasticity at the mf-GrC relay play a role in motor co-ordination and learning.

EMERGENCE OF ELEMENTARY PROPERTIES AT THE NETWORK LEVEL

It may now be asked whether evidence for mf-GrC synaptic plasticity and complex input-output dynamics can give new insight into the mechanism of cerebellar computation. In this chapter I will propose how these elementary properties may emerge at the network level.

Pattern Recognition

The demonstration of synaptic plasticity suggests that the mf-GrC relay is suitable for *pattern recognition.*[1,3] According to the neuronal network theory, this means that after learning has occurred, presentation of the original pattern should cause fast activation of a specific subset of GrCs and subsequent selection of the appropriate output. The mf-GrC relay is ideal for this task, because of the huge number of mf-GrC synapses (around 10^{11}) and the strong convergence on the PC layer. Recurrent GrC excitation (see Arbib et al.,[3,15] Boylls[16]) may improve the extraction of input patterns from noise or from an incomplete presentation, and the detection of specific features of the presented objects.[1] Recently, cerebellar network simulations showed that mf-GrC LTP greatly improves motor control through a pattern recognition mechanism (N. Schweighofer, personal communication).

Network Synchronisation

Using models of GrCs and GoCs, a simulation of the mf-GrC relay has been performed.[40] When a random mf discharge is simulated, granule cells tend to beat in synchrony. This effect probably depends on the inhibitory feed-back loop of GoCs, and reflects a general property of recurrent neuronal networks with inhibitory synapses.

Bursting in granule cells,[33] and autorhytmicity in Golgi cells,[41] may enhance the oscillating behaviour of the network. Similarly as in the olfactory bulb (see Arbib et al.,[3] Koch et al.[23]), GrC synchronisation may contribute to set the right pace to cerebellar computation.

CONCLUSIONS

A detailed experimental investigation of the mf-GrC relay has revealed unexpected properties of the input stage of the cerebellum, which may have important functional consequences.

- A single molecule, the NMDA receptor, is central to the whole computational process. NMDA receptors control both information transfer and learning, which appear therefore as two computational aspects of the same molecular mechanism.
- Synaptic plasticity in the mf-GrC relay should implement a mechanism for pattern recognition, improving cerebellar analysis of incoming signals. In this sense the cerebellum operates as a sensory system.[3]
- The demonstration of synaptic plasticity at the mf-GrC relay,[29-31] as well as at the pf-PC relay[13] and the PC-NC relay,[42] suggests that learning and memory are distributed properties of the cerebellar network. In addition, the cerebellum would influence learning and memory by regulating the function of other brain regions.[43]
- It is possible that network and neuronal properties cooperate to induce oscillating behaviours, suggesting that incoming information is processed within internal network computational states.

The way that elementary properties of neurones emerge at the network level is therefore rather complex, and its further understanding would benefit of detailed models of neuronal and network activity.

REFERENCES

1. P.S. Churchland and T.J. Sejnowski, *The Computational Brain*, MIT Press, Cambridge MA (1992).

2. C. Koch and J.L. Davies, *Large-Scale Neuronal Theories of the Brain*, MIT Press, Cambridge MA (1996).

3. M.A. Arbib, Erdi P. and J. Szentagothai, *Neural Organization: Structure, Function, and Dynamics*, MIT Press, Cambridge MA (1998).

4. R. y Cajal, *Histologie du System Nerveux de l'Homme et des Vertebres*, Vol. II, Maloine, Paris (1911).

5. J.C. Eccles, M. Ito, and J. Szentagothai, *The Cerebellum as a Neuronal Machine*, Springer Verlag, Berlin (1967).

6. M. Ito, *The Cerebellum and Neural Control*, Raven Press, New York (1984).

7. L. Luciani, *Il cervelletto: Nuovi studi di Fisiologia Normale e Pathologica*, Le Monnier, Florence (1981).

8. G. Holmes, The cerebellum of man, *Brain*, 62:1 (1939).

9. V. Braitenberg and N. Onesto, The Cerebellar Cortex as a Timing Organ: Discussion of a Hypothesis, *Proceedings of the 1st International Congress on Medical Cybernetics*, p. 1-19, Giannini, Napes (1962).

10. V. Braitenberg and H. Preissl, Why is the output of the cerebellum inhibitory? *Behav. Brain Sci.* 15:715 (1992).

11. D. Marr, A theory of cerebellar cortex, *J. Physiol.*, 202:437 (1969).

12. J.S. Albus, A theory of cerebellar function, *Math. Biosc.*, 10:25 (1971).

13. M. Ito, Long-term depression, *Annu. Rev. Neurosci.* 12:85 (1989).

14. E. De Shutter, Cerebellar long-term depression might normalize excitation of Purkinje cells: a hypothesis, *Trends in Neurosci.*, 18:291 (1995).

15. M. Arbib, C.C. Boylls, and P. Dev, Neural models of spatial perception and the control of movement. In *Cybernetics and Bionics*, eds W.D. Keidel, w. Handler, M. Spreng, p. 216-231, Munich (1974).

16. C.C. Boylls, A theory of cerebellar function with applications to locomotion: I. The physiological role of climbing fiber inputs in anterior lobe operation. Tch. rep. no. 75C 6, Amherst, MA: Computer and Science department, University of Massachussets at Amherst, 1975.
17. J.R. Bloedel, Functional heterogeneity with stuctural homogeneity: how does the cerebellum operate?, *Behav. Brain Sci.*, 15:666 (1992).
18. J-S Lou and J.R. Bloedel, Responses of sagittally aligned Purkinje cells during perturbed locomotion: relation of climbing fiber activation to simple spike modulation, *J. Neurophysiol.*, 78:1820 (1992)
19. R. Llinas and J.P. Welsh, On the cerebellum and motor learning, *Curr. Opin. Neurobiol.*, 3:958 (1993).
20. J.P. Welsh, E.J. Lang, I. Sugihara, and R. Llinas, Dynamic organization of motor control within the olivocerebellar system, *Nature*, 374:453 (1995).
21. B. Hille, *Ionic Channels of Excitable Membranes*, Sinauer Associates Inc., Sunderland Massachussets, (1992).
22. R. Llinas, The intrinsic electrophysiological properties of mammalian neurons: insights into central nervous system functions, *Science*, 242:1654 (1988).
23. C. Koch and I. Segev, *Methods in Neuronal Modeling*, MIT Press, London (1998).
24. T. McKenna, J. Davies and S.F. Zornetzer, *Single Neuron Computation*, Academic Press, London (1992).
25. T.V.P. Bliss and G.L. Collingridge, A synaptic model of memory: long-term potentiation in the hippocampus, *Nature*, 361:31 (1993).
26. D.O. Hebb, *The Organization of the Behavior*. Wiley, New York (1949).
27. E. D'Angelo, Integration and stotage of sensory-motor information: computation in the cerebellum, *Human and Machine Perception: Information Fusion*, ed. V Cantoni et al., Plenum Press, New York (1997).
28. E. D'Angelo, Synaptic excitation of individual rat cerebellar granule cells in situ: evidence for the role of NMDA receptors, *J. Physiol.*, 484:397 (1995).
29. E. D'Angelo, P. Rossi, S. Armano, and V. Taglietti, Evidence for NMDA and mGlu receptor dependent long-term potentiation of mossy fibre - granule cell transmission in rat cerebellum,. *J. Neurophysiol.*, in press (1999).
30. E. D'Angelo, P. Rossi, Integrated regulation of signal coding and plasticity by NMDA receptors at a central synapse, Neural plasticity 6:8-16 (1998).
31. P. Rossi, E. D'Angelo, and V. Taglietti, Differential long-lasting potentiation of the NMDA and non-NMDA synaptic currents induced by metabotropic and NMDA receptor coactivation in cerebellar granule cells, *Eur. J. Neurosci.*, 8:1182 (1996).
32. F. Gabbiani, J. Midtgaard, and T. Knoepfel, Synaptic interation in a model of cerebellar granule cells, *J. Neurophysiol.*, 72:999 (1994).
33. E. D'Angelo, G. De Filippi, P. Rossi, and V. Taglietti, Ionic mechanism of electroresponsiveness in cerebellar granule cells implicates the action of a persistent sodium current. *J Neurophysiol* 80, 493 (1998).
34. E. D'Angelo, G. Naldi, and P. Rossi., A compartmentl model of the cerebellar granule cell, *Cell Modelling and Cell Signalling*, ECMBM, Heidelberg (1996).
35. S.G. Bricley, S.G. Cull-Candy, and M. Farrant, Development of a tonic form of synaptic inhibition in rat cerebellar granule cells resulting from persistent activation of GABA-A receptors, *J. Physiol. (Lond)* 497, 753-759.
36. H. Monyer, N. Burnashev, D.J. Laurie, B. Sackmann, P.H. Seeburg, Developmental and regional expression in the rat brain and functional properties of four NMDA receptors. *Neuron* 12:529 (1994).
37. S.G. Cull-Candy, S.G. Brickley, C. Misra, D. Feldmeyer, A. Moniyama, and M. Farrant, NMDA receptor diversity in the cerebellum: identification of subunits contributing to functional receptors. *Neuropharmacol.* 37:1369 (1998).
38. H. Kadotani, T. Hirano, M. Masugi, K. Nakamura, K. Nakao, M. Katsuki, and S. Nakanishi. Motor discoordination results from combined gene disruption of NMDA receptor NR2A and NR2C subunits, but not from single disruption of the NR2A or NR2C subunit. *J. Neurosci.* 16:7859 (1996).
39. R. Sprengel, B. Suchanek, C. Amico, R. Brusa, N. Burnashev, A. Rozov, O. Hvalby, V. Jensen, O. Paulsen, P. Andersen, J.J. Kim, R.F. Thompson, W. Sun, L.C. Webster, S.G.N. Grant, J. Eilers, A. Konnerth, J. Li, J.O. McNamara, and P.H. Seeburg, Importance of the intracellular domain of NR2 subunits for NMDA receptor function in vivo. *Cell* 92:279 (1998).
40. R. Maex, B.P. Vos, and E. Deshutter, Dynamics of a detailed model of the granular layer of the cerebellum, *Soc. Neurosci. Abs.* 433.15 (1996).
41. J. Midtgard, Membrane properties and synaptic responses of Golgi cells and stellate cells in the turtle cerebellum *in vitro*, *J. Physiol*,. 457:329 (1992).
42. C.D. Aizenman, P.B. Manis, D.J. Linden, Polarity of long-term synaptic gain change is related to postsynaptic spike firing at a cerebellar inhibitory synapse. *Neuron* 21:827 (1998).
43. S.G. Lisberger, T.A. Pavelko, and D.M. Broussard, Neural basis for motor learning in the vestibuloocular reflex of primates. III. Computational and behavioral analysis of the sites of leaning, *J. Neurophysiol*,. 72:974 (1994).

FACE DETECTION AND CAMOUFLAGE BREAKING BY DIRECT CONVEXITY ESTIMATION*

Ariel Tankus, Yehezkel Yeshurun, and Nathan Intrator

Department of Computer Science
Tel-Aviv University
Ramat-Aviv 69978, Israel
e-mail: arielt,hezy,nin@math.tau.ac.il

INTRODUCTION

Edge detection was, so far, the core of most state of the art techniques for attentional mechanisms as well as face detection.[1,2] This excludes some recent works which utilise neural networks,[3,4] colour histograms,[5,6] or shape statistics[7,8] for face detection. Though one cannot disregard their advantages, edge maps sustain severe flaws such as: sensitivity to changes in illumination, strong effect of surrounding objects, and inability to delineate objects in a cluttered scene. We overcome these problems of edge-based approaches by a novel attentional operator which detects smooth three dimensional convex or concave objects in the image. The operator is robust to face orientation, scale, and illumination, and is capable of detecting the subject in a strongly textured background. It is employed for face detection, namely to detect the eyes and hair, from which the scale of the face can be inferred. The operator answers the above problems as a whole, demands a relatively short running time, and its robustness leads to reliable results.

Y-ARG: ATTENTIONAL OPERATOR FOR DETECTION OF CONVEX REGIONS

We next define our attentional mechanism. We refer to faces as three dimensional objects with convex and concave regions, and take advantage of this structure.

Defining the Argument of Gradient

Let us estimate the *gradient map* of image $I(x,y)$ by: $\nabla I(x,y) \approx ([D_\sigma(x)G_\sigma(y)]*I(x,y),[G_\sigma(x)D_\sigma(y)]*I(x,y))$, where $G_\sigma(t)$ is the one dimensional

*This chapter is a modified version of a paper in PATTERN RECOGNITION LETTERS, Vol 18(9), 1997, pp 913-922, with permission from Elsevier Science.

Gaussian with zero mean and standard deviation σ, and $D_\sigma(t)$ is the derivative of that Gaussian. We turn the Cartesian representation of the intensity gradient into a *polar* representation. The *argument* (also denoted "phase", and usually marked by $\theta(x,y)$), is defined by:

$$\theta(x,y) = \arg(\nabla I(x,y)) = \arctan(\frac{\partial}{\partial y} I(x,y), \frac{\partial}{\partial x} I(x,y)) \tag{1}$$

where the two dimensional arc tangent is defined by:

$$\arctan(y,x) = \begin{cases} \arctan(\frac{y}{x}), & \text{if } x \geq 0 \\ \arctan(\frac{y}{x}) + \pi, & \text{if } x < 0,\ y \geq 0 \\ \arctan(\frac{y}{x}) - \pi, & \text{if } x < 0,\ y < 0 \end{cases} \tag{2}$$

and the one dimensional arctan(t) denotes the inverse function of tan(t) so that: $\arctan(t):[-\infty,\infty]\rightarrow[-\pi/2,\pi/2]$. While the term "phase" is widely used in the literature,[9,10] we use it in a completely different manner: the term "phase" in this chapter refers to the argument of the intensity gradient.

The attentional mechanism is simply the derivative of the argument map with respect to the y-direction: $\partial/\partial y\ \theta(x,y) \approx [G_\sigma(x)D_\sigma(y)]*\theta(x,y)$. We denote $\partial/\partial y\ \theta(x,y)$ as *Y-Arg*.

Mathematical Formulation of Y-Arg Reaction to Paraboloids

The projection of concave and convex objects can be estimated by paraboloids, since paraboloids are arbitrarily curved surfaces.[11] Our mathematical formulation refers to a general paraboloid of the form: $f(x,y) = a(x-\varepsilon)^2 + b(y-\eta)^2$, where $a>0$, $b>0$ are constants, and (ε, η) is the center of the paraboloid. The first order derivatives of the paraboloid are: $\partial/\partial x\ f(x,y) = 2a(x-\varepsilon)$ and $\partial/\partial y\ f(x,y) = 2b(y-\eta)$.The gradient argument is therefore: $\theta(x,y) = \arctan(b(y-\eta), a(x-\varepsilon))$. Deriving it with respect to y yields: $\partial/\partial y\ \theta(x,y) = [ab(x-\varepsilon)]/[a^2(x-\varepsilon)^2+b^2(y-\eta)^2]$. However, this derivative exists in the whole plane except for the ray: $\{(x,y) \mid y=\eta \text{ and } x\leq\varepsilon\}$. At this ray, $\theta(x,y)$ has a first order discontinuity (in the y–direction), so its derivative there tends to infinity. The fact that for a paraboloid, $\partial/\partial y\ \theta(x,y) \rightarrow \infty$ at the negative ray of the x-axis, while continuous at the rest of the plane can be clearly seen in Figure 1(c) (we define our coordinate system at the horizontal and vertical axes of the sphere).

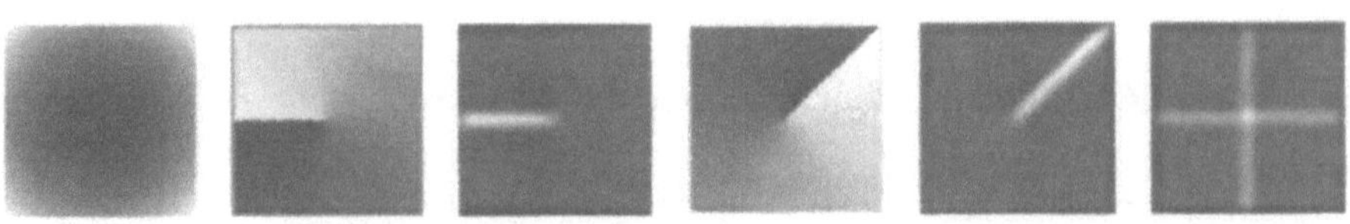

(a) (b) (c) (d) (e) (f)

Figure 1. (a) The spheric gray-levels: $I(x,y) = 10x^2 + 10y^2$; (b) the argument of gradient of (a). The discontinuity ray is at 180° from the positive x-axis; (c) derivation of (b) in 90°; (d) rotation of (b), so that the discontinuity ray is at 45° from the positive x-axis; (e) derivation of (d) in 315°; (f) response of D-Arg, the isotropic operator with rotation angles α=0°, 90°, 180°, 270°.

D-Arg: The Isotropic Variant

We also define an isotropic variant of Y-Arg, whose reaction is to all axes of the paraboloid, rather than merely the negative part of the x-axis. We do so by rotating the gradient argument by:

$$\theta_\alpha(x,y)=\begin{cases}\theta(x,y)+(\pi-\alpha), & \text{if } \theta(x,y)+(\pi-\alpha)\le\pi\\ \theta(x,y)+(\pi-\alpha)-2\pi, & \text{if } \theta(x,y)+(\pi-\alpha)>\pi\end{cases} \tag{3}$$

so the ray of discontinuity of Y-Arg is transformed to a ray from the origin forming an angle of α radians with the positive part of the x-axis. We then derive the rotated argument of gradient in the direction: α-$\pi/2$ (or: $\alpha+\pi/2$), to get the response to the ray of discontinuity (see Fig. 1(d),(e)). Repeating this rotation with angles: $\alpha = 0°, 90°, 180°, 270°$ and summing their responses result in an isotropic operator called: *D-Arg* (see Fig. 1(f)). It is evident that D-Arg is more general than Y-Arg, but as we shall show, Y–Arg is effective and robust for face detection.

FEATURES OF Y-ARG

Two dimensional objects of constant albedo form a linear gray-level function, and are usually of no interest (for example, walls). It can be easily shown, that the Y-Arg attentional mechanism has zero response to planar objects. In addition, it can be shown that the response of Y-Arg to edges of planar objects is finite, and is therefore smaller than its response to paraboloids. Another provable characteristic of Y-Arg is its linear dependence on scale. We now present several invariants of Y-Arg, followed by a practical discussion and demonstration from real-life scenes.

Theorem 1 *Let $f(x,y)$ [the original gray-level function] be a derivable function at each pixel (x_0, y_0) with respect to x and y.*
Let $T(z)$ [the transform] be a function derivable at point $z_0 = f(x_0,y_0)$, whose derivative there is positive in the strong sense. Denote the composite function: $g(x,y) = T(f(x,y))$ [the transformed grey-level function].
The y-derivatives of the gradient arguments of $f(x,y)$ and $g(x,y)$ at point (x_0,y_0) are identical:

$$\frac{\partial\theta_g(x_0,y_0)}{\partial y}=\frac{\partial\theta_f(x_0,y_0)}{\partial y} \tag{4}$$

Proof: By the *chain rule*, the composite function: $g(x,y) = T(f(x,y))$ is derivable with respect to both x and y at point (x_0, y_0), and its derivatives are:

$$\begin{aligned} g_x(x_0,y_0)&=T'(f(x_0,y_0))f_x(x_0,y_0)\\ g_y(x_0,y_0)&=T'(f(x_0,y_0))f_y(x_0,y_0)\end{aligned} \tag{5}$$

We denote $f^0=f(x_0,y_0)$, $f_x^0=f_x(x_0,y_0)$, $f_y^0=f_y(x_0,y_0)$. The argument of gradient at point (x_0,y_0) can be written as:

$$\theta_g(x_0,y_0)=\arctan\left(T'(f^0)f_y^0,T'(f^0)f_x^0\right) \tag{6}$$

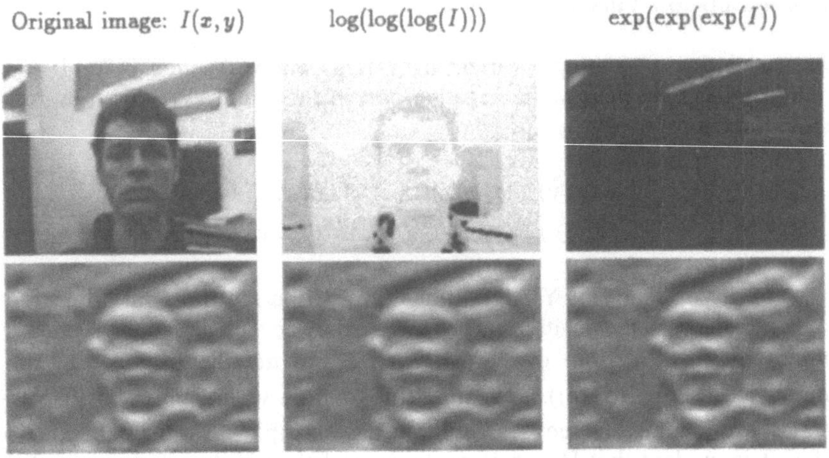

Figure 2. Top row: The original image I(x,y) is compared to log(log(log(I(x,y)))) and exp(exp(exp(I(x,y)))). Y-Arg is invariant under log and exp. Bottom row: Y-Arg. Similarity among Y-Arg of original image and Y–Args of transformed images is obvious.

Since we have required that $T'(f^0) > 0$, the point $\left(T'(f^0)f_x^0, T'(f^0)f_y^0\right)$ lies in the same quarter of the plane as point (f_x^0, f_y^0). It follows that:

$$\theta_g(x_0, y_0) = \arctan\left(T'(f^0)f_y^0, T'(f^0)f_x^0\right) = \arctan(f_y^0, f_x^0) = \theta_f(x_0, y_0) \tag{7}$$

The last equation states that the phase of the gradient is invariant under the transformation T. Deriving the gradient argument with respect to y preserves this invariance:

$$\frac{\partial\theta_g(x_0, y_0)}{\partial y} = \frac{\partial\theta_f(x_0, y_0)}{\partial y} \tag{8}$$

Let us rephrase theorem 1 in the following manner:

Y-Arg is invariant under any derivable monotonically increasing (in the strong sense) transformation of the grey-level function.

The practical meaning of the theorem is that Y-Arg is invariant, for example, under linear transformations, positive powers (where $f(x,y) > 0$), logarithm, and exponent. Y-Arg is also invariant under linear combinations (with positive coefficients) and compositions of these functions, since such combinations are also derivable and strongly monotonically increasing. The functions mentioned above and their combinations are common in image processing for lighting improvement. This implies that Y-Arg is invariant under a large variety of lighting conditions. Figure 2 demonstrates Y-Arg invariance to log(log(log(z))) and exp(exp(exp(z))) in a real-life scene.

In view of Y-Arg invariants, the suggested model is not only a paraboloidal grey-levels detector, but also a detector of any derivable (strongly) monotonically increasing transformation of paraboloids. This makes Y-Arg particularly attractive for usage in various scenes in which the environment is unknown before hand.

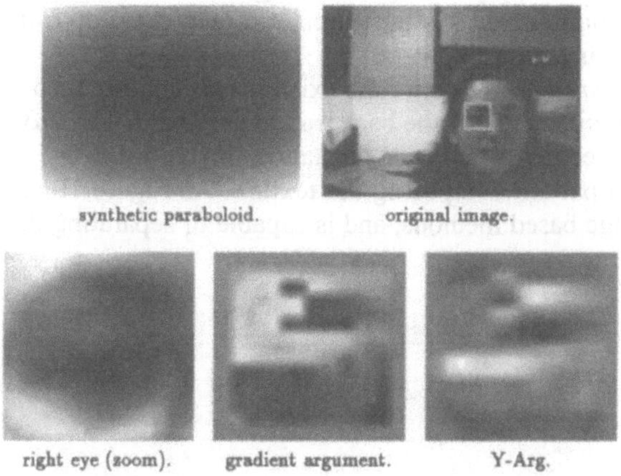

Figure 3. The eye exhibits strong similarity to the artificial paraboloidal grey-levels: $I(x,y) = 10x^2 + 30y^2$. The gradient argument of the eye is similar to that in Figure 1. A clear response of Y-Arg at the negative part of the x-axis is observed.

FACE DETECTION USING Y-ARG

Approximation by Paraboloids

One of the underlying ideas of the theoretical model is the estimation of the grey-levels describing convex and concave objects, in our case - the eyes and hair, using paraboloids or a derivable monotonically increasing transformation (in the strong sense) of paraboloids. Figure 3 shows such a synthetic paraboloid along with a magnified eye. The eye grey-levels are similar to those of a paraboloid. The Y-Arg of the eye strongly reacts to the x-axis; this behaviour resembles that of Y-Arg of paraboloids.

Figures 4, 5, 6, demonstrates the robustness to three factors: illumination direction, scale, and orientation of the head, respectively. Mirrored auto-correlation serves to detect the face, i.e., choose the window with the best cross correlation between left and right halves (mirroring one of them) among all possible window positions; the window is of the same height as the image. Y-Arg robustness to illumination, scale, and orientation is mainly due to the fact that $\partial/\partial y\ \theta(x,y) \rightarrow \infty$ for paraboloids, which is a very stable feature. The areas of strong Y-Arg response enable the heuristic detection of the face scale.

Superiority of Y-Arg on Edge Detection

In this section, we briefly delineate the results of an extensive comparison between Y–Arg and the edge map (taken as gradient modulus). Following the operation of each method, mirrored auto-correlation attempts to detect the face:

• *reaction to 3D objects*: Y-Arg detects 3D objects, so the Y-Arg of a smiley (2D object) is relatively low, as opposed to edge methods (see Fig. 7);

• *insensitivity to strong edges*: Sharp colour changes are likely to appear due to different object colours (or albedo) (see Fig. 8). These variations lead to strong edges, which distract edge-based methods from the subject. Y-Arg does not react strongly to sharp changes, but rather, to gradual changes of intensity of the kind exhibited by the eyes and hair;[12]

• *robustness to lighting*: in Figure 9(a), the background is better lit than the subject. Y-Arg detects the subject, while edge methods, the background. The improvement of the image by the logarithm (see Fig. 9(b)) makes edge-based methods too detect the subject.

Note, that the decision to apply this specific (log) function was made by a human. Y-Arg robustness to illumination releases the automatic face detector from the need to decide which illumination it is facing;

• *stability in textured background*: the existence of texture in an image makes the task of discriminating the subject from the background very hard. The difficulties emanate from the large amount of edges covering a substantial image area, and the periodicity of the (usually symmetric) pattern composing the texture. As Figure 10 shows, Y-Arg is much more robust than edge based methods, and is capable of separating the face from dominant textures.

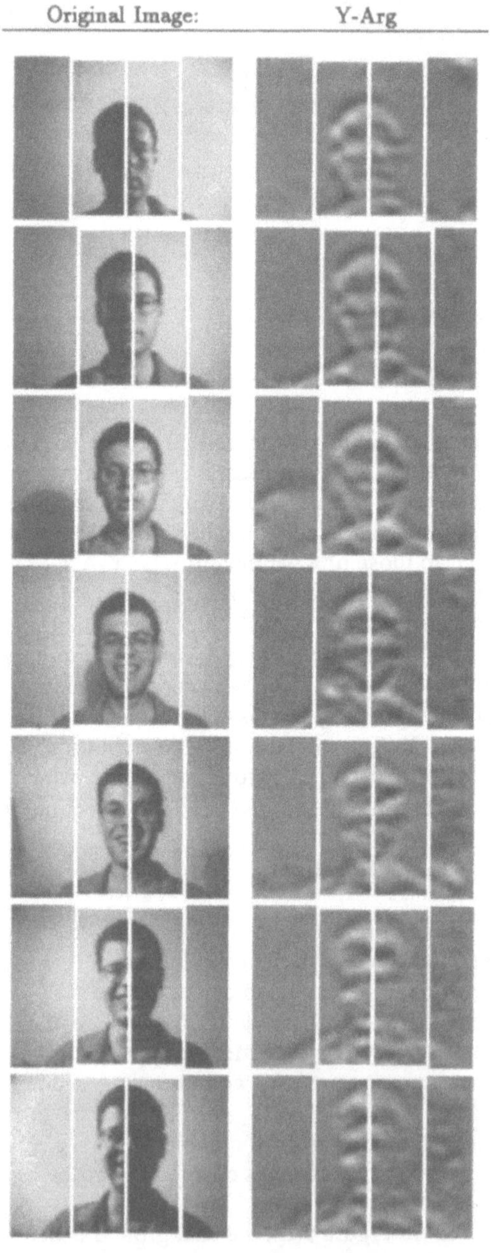

Figure 4. Robustness to lighting. Illumination comes from a single point light source. Each row relates to the corresponding azimuth: 90°, 60°, 30°, 0°, -30°, -60°, -90°. Detection by mirrored auto-correlation is marked.

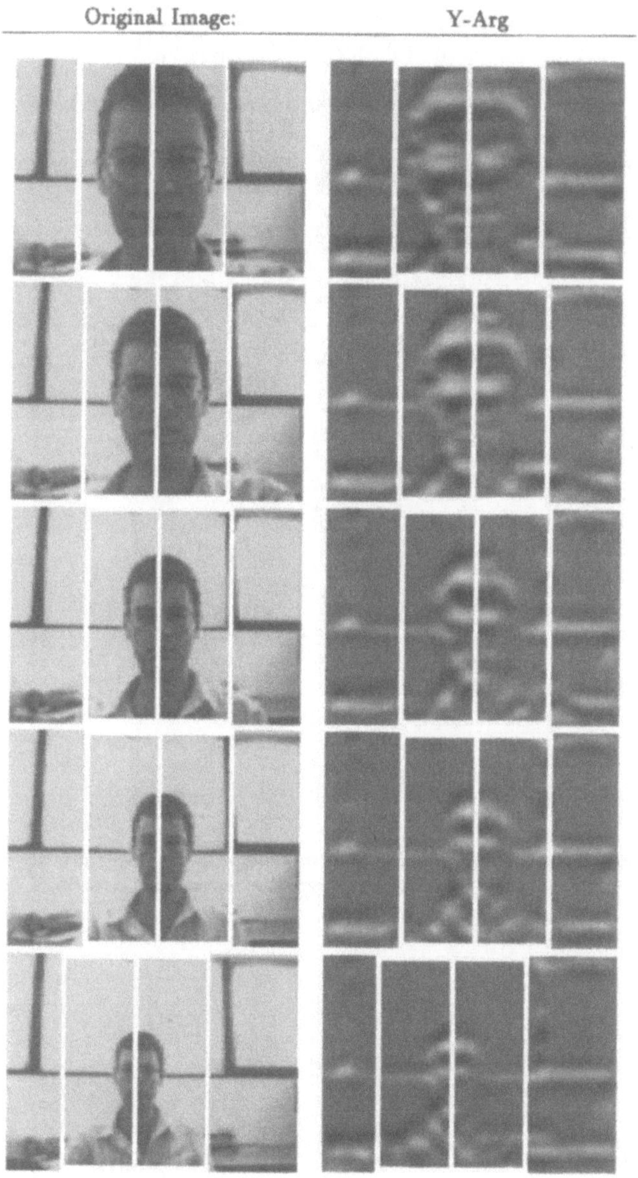

Figure 5. Robustness to scale. Several scales of the face. The Y-Arg image strongly reacts to the eyes and hair regions, regardless of the scale. Largest face is about 6 times larger than smallest face. Detection by mirrored auto-correlation is marked.

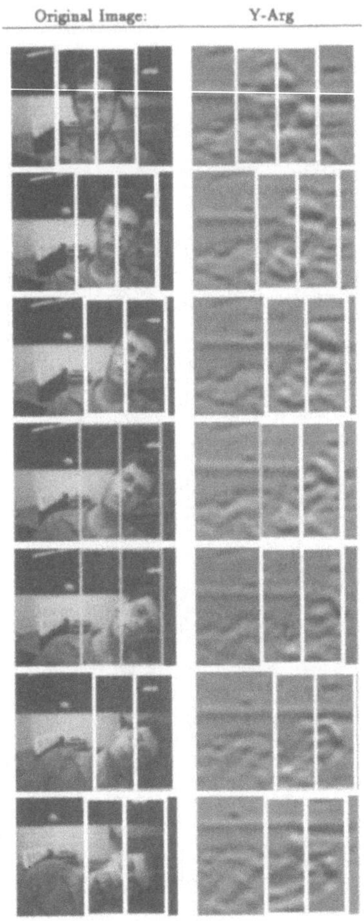

Figure 6. Robustness to orientation. Capability to detect oblique faces. Y-Arg strongly reacts to the eyes and hair, even though the face is slanted. Detection by mirrored auto-correlation is marked.

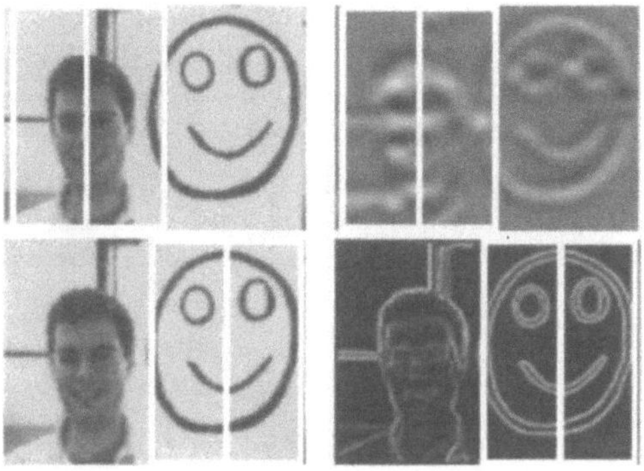

Figure 7. A face vs. smiley. Edge-based methods locate the larger object, which is the flat smiley. Y-Arg detects the three dimensional face, although it is smaller.

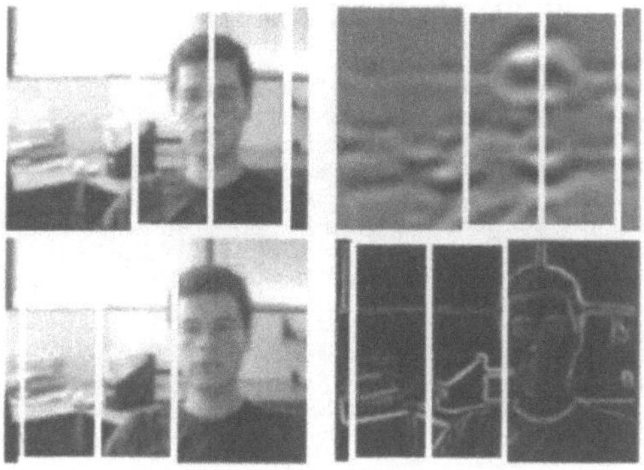

Figure 8. Small objects with strong edges divert edge-based methods from the real subject. Y-Arg reacts to gradual variations rather than to sharp colour changes.

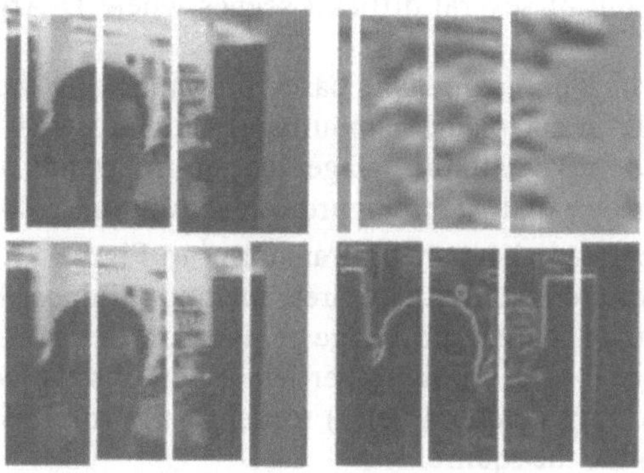

(a)

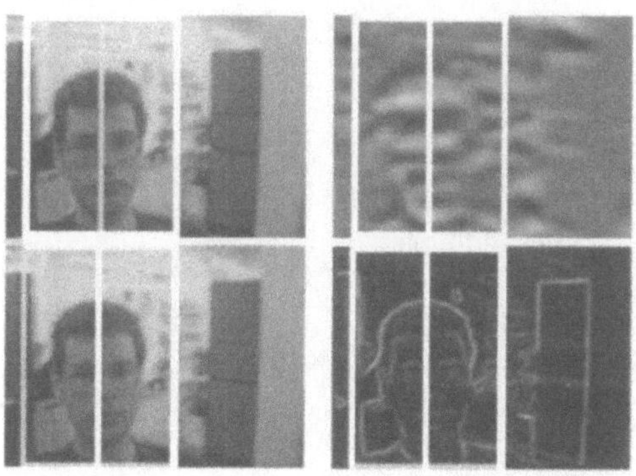

(b)

Figure 9. (a) The background is better lit than the face. Y-Arg is capable of detecting the face despite the poor illumination. (b) Illumination improved by applying logarithm. Edge methods as well as Y-Arg now detect the subject.

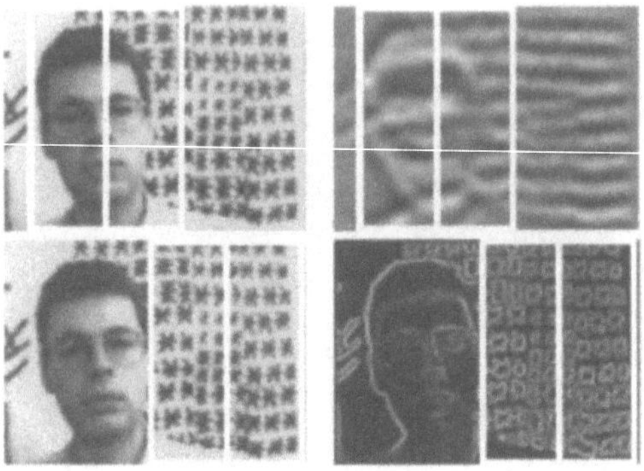

Figure 10. A texture of asterisks. Facial edges look negligible near texture edges. In the Y-Arg map, facial regions attain higher values than textural areas.

CAMOUFLAGE BREAKING

The stability of D-Arg under various conditions (illumination, scale, orientation, texture) makes it suitable for camouflage breaking. This section demonstrates one camouflage example out of several difficult scenes where D-Arg succeeded in breaking camouflage.

The camouflage of many animals is based on mimicking the environment. Both the colour and the texture of fur of the camouflaged animal fit its habitat. Figure 11 is an example of animal camouflage. In this image, the squirrel stands on a ground covered with leaves, under the shades of a tree. Figure 12 shows the response of an edge-based attentional operator, the radial symmetry transform[2] to Figure 11. The strong camouflage yields a strong response to a large image area, making it impossible to detect the squirrel from the symmetry map. No single image region can be isolated using the symmetry transform in this case. D-Arg, on the other hand, exploits the convexity of the squirrel (especially - its uniformly coloured belly) to detect the squirrel. D-Arg has two strong peaks, both correspond to the squirrel.

Figure 11. A squirrel on the background of leaves covering the ground, under the shades of a tree. A human viewer is usually disguised by the camouflage.

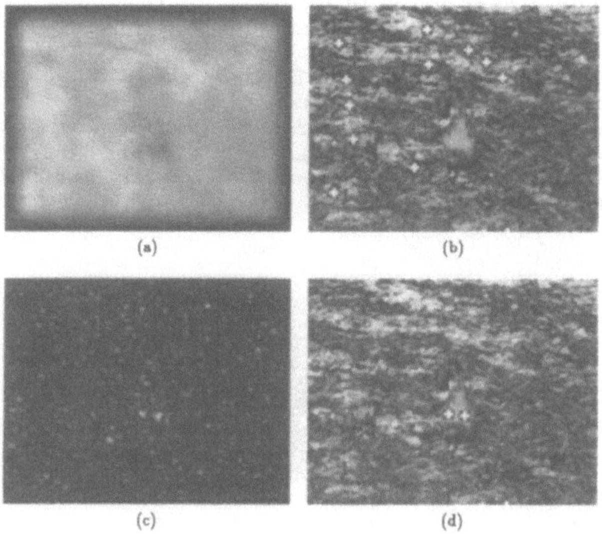

Figure 12. Detection of the squirrel of Figure 11. (a) Radial symmetry map (radius=30). (b) Detection by radial symmetry (90% threshold). (c) D-Arg^2. (d) Detection by D-Arg (only 70% threshold are required for detection).

CONCLUSIONS

We introduce a novel attentional operator (Y-Arg) for detection of regions emanating from smooth convex or concave three dimensional objects. We use it to detect the eyes and hair, and thus, the face. Y-Arg is proved invariant under any derivable (strongly) monotonically increasing transformation of the image grey-levels, which practically means robustness to illumination changes. Robustness to orientation and scale is also described. The operator is *not* based on edge maps, and thus free of their flaws (e.g., Y-Arg is robust in dominant textures). An extensive comparison with edge-based methods is depicted. Finally, we have demonstrated the usage of D-Arg (an isotropic variant of Y-Arg) for camouflage breaking in highly textured environments.

REFERENCES

1. A. Jacquin and A. Eleftheriadis, *Automatic Location Tracking of Faces and Facial Features in Video Sequences*, Proc. of the Intl. Workshop on Auto. Face- and Gesture-Recognition, Zurich (1995).
2. D. Reisfeld, H. Wolfson, and Y. Yeshurun, *Context Free Operators: The Generalized Symmetry Transform*, Intl. Journal of Computer Vision, Vol.14, (1995).
3. K-K. Sung and T. Poggio, *Example-Based Learning for View-Based Human Face Detection*, in Proc. from Image Understanding Workshop, Monterey, CA, Nov.1994.
4. H.A. Rowley, S. Baluja, and T. Kanade, *Human Face Detection in Visual Scenes*, will appear in: Advances in Neural Information Processing Systems 8.
5. Y. Dai and Y. Nakano, *Extraction of Facial Images from Complex Background Using Color Information and SGLD Matrices*, Proc. of the Intl. Workshop on Auto. Face- and Gesture-Recognition, Zurich (1995).
6. B. Schiele and A. Waibel, *Gaze Tracking Based on Face-Color*, Proc. of the Intl. Workshop on Auto. Face- and Gesture-Recognition, Zurich (1995).
7. M.C. Burl, T.K. Leung, and P. Perona, *Face Localization via Shape Statistics*, Proc. of the Intl. Workshop on Auto. Face- and Gesture-Recognition, Zurich (1995).
8. B. Moghaddam and A. Pentland, *Maximum Likelihood Detection of Face and Hands*, Proc. of the Intl. Workshop on Auto. Face- and Gesture-Recognition, Zurich (1995).

9. S. Fischer and J. Bigün, *Texture Boundary Tracking with Gabor Phase*, G. Borgefors (Ed.), Proc. of the 9th Scandinavian Conf. on Image Analysis, pp. 877-884, Uppsala (1995).
10. D.J. Fleet and A.D. Jepson, *Computation of Component Image Velocity from Local Phase Information* Intl. J. Comp.\Vision 5, pp.77-104, (1990).
11. S.W. Zucker, M.S. Langer, L.A. Iverson, and P. Breton, *Shading Flows and Scenel Bundles: A New Approach to Shape from Shading*, ECCV '92, Springer-Verlag.
12. H.P. Graf, T. Chen, E. Petajan, and E. Cosatto, *Locating Faces and Facial Parts*, Proc. of the Intl. Workshop on Auto. Face- and Gesture-Recognition, Zurich (1995).
13. M. Bichsel ed., *Proceedings of the First International Workshop on Automatic Face- and Gesture-Recognition*, Zurich, Switzerland, June 1995.

COPING WITH UNCERTAINTY: THE SELF-ORGANISATION OF SOCIAL SYSTEMS*

Günter Küppers

Institute for Science and Technology Studies
University of Bielefeld
P.O. Box 10 01 31
33501 Bielefeld - Germany

INTRODUCTION

Nowadays, "self-organisation" is a broad term applied to a range of concepts that have one thing in common despite various names such as synergetics, autopoiesis, dissipative structures or self-referring systems: the attempt to describe and grasp the behaviour of complex, dynamic systems. In physics, for example, the concern is to explain structure-building processes in hydrodynamic convection flows or the coherent behaviour of light emissions in the laser. Chemistry studies the origins of spatial and/or temporal structures in chemical reactions. On the borders between chemistry and biology, scientists are studying the emergence and development of highly complex organic molecules and trying to understand how biological information emerged in a prebiotic world. From neurophysiology to ecology, biology is studying phenomena of ontogenesis and phylogenesis in order to grasp how the complex can emerge from the simple.

The social sciences also exhibit efforts to model the emergence of social order through such approaches. The primary issue in such studies is how far patterns of social behaviour can be explained through the orientation of the individual towards the respective other. As valuable as such studies may be, they fall short when it comes to the emergence of social rules in general and their impact on social behaviour.

This issue of the mutual dependencies between behavioural rules and behaviour has been a focus of sociological interest since the discipline was founded as the science of society by Auguste Comte (1798-1857) at the beginning of the last century. New theoretical approaches were applied repeatedly in a search for those characteristic features that could be used to describe society as a whole as well as the social behaviour of its members. From the very beginning, these two aspects of the social were considered to belong together: social structures as the forms and rules of coexistence and social processes as the social behaviour of individuals. However, no theoretical tools were available for an adequate description of the mutual dependencies between structure and

*Translated by Jonathan Harrow.

Human and Machine Perception 2: Emergence, Attention, and Creativity,
edited by Cantoni *et al.* Kluwer Academic / Plenum Publishers, New York, 1999.

process. As a result, the starting point for most sociological theories was always only the one half of this pair of concepts: *either* society *or* the individual. In each case, the other side then became a consequence of this initial choice: those who had chosen the individual modelled society as a superimposition of single, socially active individuals; alternatively, those who chose society tried to explain social behaviour as the inevitable effect of society. Table 1 sketches this discourse on the relation between the individual and society.*

Table 1. Theoretical Approaches in Sociology.

Theory	Starting point	Research focus	Theoretical categories	Individual /Society
Exchange theories	Exchange processes between individuals.	How do exchange relations structure the actions of individuals and the distribution of resources?	Cost/utility, rationality paradigm.	Statements on social facts have to be traced back to statements over individuals.
Interactionism	The interaction of individuals.	How are social relations possible; how do personal and social identity relate?	Expectations, roles, frameworks, intersubjectivity, everyday life.	Society arises through the symbolic forms of interaction among its members.
Functionalism	Societal functions.	How do function and social structure relate?	Functional equivalents, structures, integration and differentiation.	Society is defined over functions. Individuals have to subordinate themselves to this functionality.
Structuralism	Society as social structures.	How do social structures control the social behaviour of individuals?	Culture, rites, institutions, risky situations.	Societal differentiation shapes individuals as members of certain classes.
Systems theory	"There are (social) systems".	How do systems organise their survival?	Autopoiesis, observation, re-entry, structural coupling.	The subsystems of society are forms of communication, and communication is the basic element of the subsystems.

The table shows that none of the classical approaches succeeded in modelling the mutual constitution of society and individual, of social processes and social structures. Only sociological systems theory assumes a circular relationship between structure and process, between system and system-specific processes. Its basic differentiation no longer addresses the classic issue of the relationship between the parts and the whole. Instead, it initially distinguishes between system and environment. The decisive question is how a system *itself*, that is, through its own operations, emerges from its environment and makes this emergence permanent. This exclusion of the environment makes such systems "operationally closed"; in other words, they become autonomous, and their structures are exclusively an outcome of their own operations. This construction makes it possible to relate system structure to system dynamics, and it brings us closer to answering the classic

*Overviews of the various concepts in sociology can be found in, for example, Turner[1] or Kiss.[2]

question in sociology regarding the mutual dependencies between structure and process, between society and individual behaviour.

As a result, *operational closure* has become the central concept in sociological systems theory. As a social mechanism, it is responsible not only for the emergence of the social as an autonomous phenomenal domain but also its organisation through the self-reference of social processes. *Niklas Luhmann*, the founder of sociological systems theory, views communications as the basic operations in social systems. He constructs operational closure through the special nature of communication: it is bound "to the discernable breadth of the . . . operations".[3] One talks . . . "about operationally closed systems, meaning that such systems differ from their environment through the type of operations they have".[4] Communication can occur only within communication. However, communication is not the medium that links together the social activities of individual protagonists, thus forming them into members of a social system. Those who communicate do not belong to the social system; they are located in its environment.

The sociological systems theory developed by Luhmann assumes the existence of the social as an autonomous phenomenal domain. "There are systems" is Luhmann's ontological credo. Their existence is tied to social operations (communication) that establish the system-environment differentiation through their particularity while achieving the self-reproduction of the system through their internal dynamics. The one and only goal of these dynamics is to reduce environmental complexity. Luhmann describes the demands this places on communication; what qualities communication has to possess as a social operation; what can threaten it; and what may stabilise it. This highly detailed "communication theory" is only one aspect of his total concept for a new theory of society. It has only loose ties with systems theory, the second pillar of Luhmann's theoretical edifice.* For example, he does not tell us what compels communication to communicate or why it communicates in the way it does in science and in the way it does in politics; in one way in business, in another way in love. How system structure (institutions) and system operations (communication) are linked together and in what ways they are mutually dependent remains unclear, because Luhmann does not specify the social mechanisms that are responsible for such a closure of structure formation and "process control". However, it is precisely these mechanisms that are the decisive determinants of the dynamics of the social. Figure 1 depicts this circular link between structure and process.

The two mechanisms *construction* and *evaluation* describe how social processes build up structures, and how these structures determine the social processes. The goal of this paper is to develop an appropriate model of such a circular closure of process and structure for the social, to specify the mechanisms of this closure in more detail, and thus to explain the emergence of social order. The theory of self-organisation will make a major

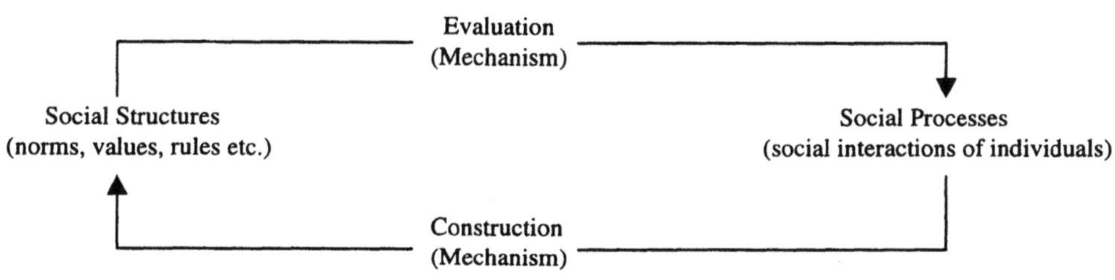

Figure 1. Circular constitution of social processes and structures.

*The third pillar is his theory of evolution.

contribution to this programme because it describes the emergence of order in natural processes through the circular closure of cause and effect. Therefore, the following will start with the theory of self-organisation, but not as an abstract mathematical theory, but in a form that permits its generalisation to the domain of social phenomena without paying the price of trivialisation inherent to a mathematisation.* What such a generalisation might look like is the topic of the second section. The final section will apply such a model of social self-organisation to innovation processes.

SELF-ORGANISATION: COUPLING ENVIRONMENTAL STRUCTURE AND SYSTEMS DYNAMICS

To a large extent, the world in which we live is well-ordered. Sometimes this order is simple; at other times, it is more complex and is not apparent at first glance. Where does this order come from? Is it imposed externally, as in my bookcase where the books are organised according to content, or has it emerged by itself as in the stripes of the zebra?** However, how can order emerge by itself? How can a crowd of people order themselves when they have no idea how to do this? There are two ways to answer such questions: either the order of things is a property of the things themselves, or mechanisms must exist that select the one option we see from the variety of possible options: stripes on the zebra and not spots. Science searched for these mechanisms for a long time, and a formally satisfying model for a solution was found only in the 1960s. This theory of self-organisation shows how the individual option is selected from the many possibilities under concrete conditions, for example, precisely the one distribution in stripes from the potential distribution of the colours black and white on a zebra's hide.

In many domains, the theory of self-organisation is now available in precise mathematical terms. For many examples from science, it shows which mechanism is responsible for the emergence of organisation: the circular coupling of cause and effect. Such a coupling is arranged constructively in experimental designs or in technical instruments. In the laser, for example, the coupling between the field of light and matter is stimulated by selecting appropriate materials, and this creates the experimental precondition for the phases of the light waves emitted to oscillate in synchrony.[6] In the thermostat, to take another example, a temperature probe regulates the operation of a heating system through a switch, but one could also say that the heating system regulates the switch through the temperature probe. This mutual dependence between cause and effect means that the thermostat is a technical device that regulates itself. It ensures that the temperature in a room remains constant regardless of the temperature outside.

Circular causality is the hallmark of self-organisation. As a property of nature, and not as a technical construction, it is to be found wherever non-equilibria strive towards

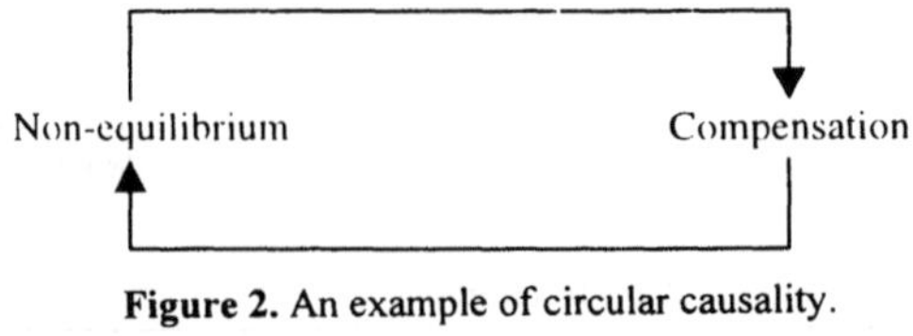

Figure 2. An example of circular causality.

*Such a conceptualisation of social self-organisation has to be performed before one can start to look for ways to find a mathematical description.

**Naturally, one could object that the information on this pattern of stripes is coded in the genes. However, recent studies have shown that pattern formation in biology is the outcome of a self-organisation process.[5]

equilibrium. This is because the compensation as an effect changes the non-equilibrium as its cause, and the changed cause, in turn, produces a changed effect (see Fig. 2).

One example of non-equilibrium is a difference in temperature within a fluid. Under certain conditions, this temperature difference will set in motion a flow that changes the temperature profile in the fluid. This closed circle between non-equilibrium and compensation will always organise itself when non-equilibrium and compensation continue to modify each other until they become mutually dependent: a certain form of compensation reproduces precisely the non-equilibrium that has this form of compensation as its consequence.

Such a form of compensation that reproduces its own cause can be observed as an order. Order is a certain pattern in the flow of compensation processes. Bérnard convection is certainly the best-known example for the emergence of organisation in fluid layers through a difference in temperature.[7]

More abstractly, and in the terminology of systems theory, the circular closure of cause and effect separates the part of the world that is embraced by this closure. This leads to the difference between system and environment: the system as a dynamic set of effects of non-equilibrium and compensation whose states are determined solely by this internal dynamics; and the environment as the rest of the world that provides the necessary resources for the maintenance of the non-equilibrium in the system. This conceptualisation implies that there are locally bounded non-equilibria in the world that are prevented from "globalising" themselves through corresponding structures in the world.* For example, the earth's oceans and skies have different temperature profiles that lead to specific forms of order such as ocean currents or cloud formations. Because of the operational closure of cause and effect in the system, the environment can only disturb the system. It cannot control the states of the system. In contrast to classic systems theory, it is no longer the environment that specifies the system, but the system that specifies the environment.

The closure of cause and effect, which generates the separation of system and environment, is also responsible for the formation of structure within the system (see Fig. 3). This is because closure permits only very specific mutual dependencies of cause and effect: only those that mutually reproduce each other. A specific cause C_0 produces a specific effect E_0 which reproduces its cause. The reproducing effect E_0 can be observed as a state of order. This mutual self-reproduction of cause and effect can occur in numerous ways within complex (non-linear) systems. For instance: which form of compensation emerges depends completely on the type of non-equilibrium at the edge of the system, and this form, in turn, is determined by the structures in the environment. This is because these structures provide the necessary resources for maintaining the non-equilibrium. In this sense, the organisation of the system reflects not only the structure of the environment but also the internal dynamics of the system. This aspect will be considered in more detail when discussing the example of innovation networks. The circular coupling of cause and effect mediates between environmental structure and system dynamics, thereby making the organisation of the system independent from changes in the environment. Systems can "survive" in a changing environment and thus retain their identity. Only dramatic

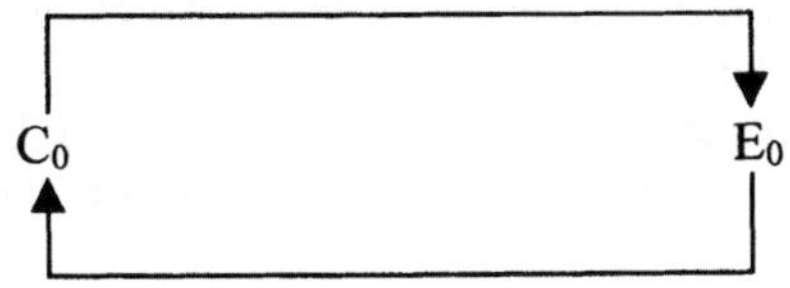

Figure 3. The reproduction of cause and effect.

*Heinz von Foerster[8] already pointed out that self-organisation is possible only in a structured environment.

fluctuations in the environment may go beyond their ability to adapt. Adaptation is therefore not an accomplishment of the environment, as in Darwin, but an accomplishment of the system.

THE SELF-ORGANISATION OF THE SOCIAL

If we want to generalise this concept of self-organisation to social processes and explain the emergence of social organisation, we could, in analogy to the natural sciences, search for non-equilibria in the distribution of material or non-material goods in the social world. It would be even better, because more general, to keep an eye open for social mechanisms that link social processes with their causes and, in this way, accomplish the emergence and reproduction of social systems. The triggering of social activities through the perception of social uncertainties* can be proposed as a mechanism for linking cause and effect. This coupling of perceived uncertainty and coping with uncertainty possesses exactly the right architecture of circular causality and therefore seems to be appropriate for an adequate description of phenomena of social self-organisation.

Therefore, we shall start with a group of individuals in a state in which regulative effects of the social can be disregarded. In such a "pre-social" world, individual protagonists simply pursue their individual interests.[8] If several protagonists compete for scarce resources in an effort to increase individual utilities, mutual obstruction soon occurs. If they all want the same, they enter into a confrontation through which the potential loss may be far greater than the anticipated gain. In the model from game theory that has become known as the Prisoner's Dilemma, two prisoners may reduce the length of their sentences if they reach an agreement. However, if one prisoner breaks this agreement, he will be released immediately, whereas the other will receive a much higher punishment. If both break the agreement, neither will gain anything.

Confrontation is risky. Nobody knows the strategies of the other or has any means of discerning them. Each acts under uncertainty, yet has to anticipate that anything may happen, and the only solution seems to be to eliminate the other. However, even then, success is uncertain, and both adversaries face the risk of being eliminated themselves. The solution to this dilemma is cooperation. By agreeing on social rules of cooperation (consensus, agreements, contracts, laws, moral obligations), the uncertainty of confrontation can be reduced sufficiently to allow any activities that pursue personal interests and are oriented towards subjective utility to become possible in the first place. These particularly involve the three sets of rules governing economic *exchange*, sanctions in *hierarchies* and *solidarity* in groups. It is obvious that these encourage order and make the risks involved in actions more transparent.

The transition from unregulated juxtaposition or conflict to regulated cooperation marks the ideal case of the emergence of the social as an autonomous phenomenal domain. The social is defined here as the set of all rules that regulate social coexistence either implicitly or explicitly, and in such a way that social conflicts can be avoided as far as possible as sources of uncertainty. However, even in a social world that has already emerged, the risks of cooperation are not transparent. This is because rules of cooperative behaviour hold only for the specific problem state for which they were developed. Rules that regulate cooperation on the roads are useless for regulating intimate relationships. The uncertainty of the unregulated will never disappear. Rules can only reduce it to a socially acceptable level in a concrete case and for a certain length of time, and thus transform uncertainty into a state of relative certainty. As this shows, the problem of regulating social cooperation never ends.

---

*Uncertainty is used here to qualify a state that is perceived as comparatively unregulated and thus as uncertain by the individuals acting within it.

If rules of cooperation prove their worth in a domain, then they may become institutionalised. Contracts are made, standing rules are drawn up and institutions are assigned the task of prosecuting rule violations and imposing sanctions. And because social rules always have to be interpreted first and do not determine social behaviour, there is also enough uncertainty within the institutional structures of a society for social coping with uncertainty to be viewed as a driving force of the social.

The social perception of uncertainty and the social activities for coping with it provide us with the social equivalent of circular causality in natural processes. Whereas natural systems form structures, social structures generate rules; in other words, they produce information: coping with the uncertainty of the unregulated leads to the formation of rules, and the rules reduce the uncertainty. The circular closure of cause and effect is the mechanism for the emergence of order in the social world as well: a certain form of uncertainty triggers a specific way of coping with it that modifies this uncertainty until it becomes socially acceptable and therefore conveys a feeling of social certainty.

The forms of coping with uncertainty in society can be observed as social order. They consist of action rules that consolidate conservatively into the various institutions of society such as legal bodies, business organisations, networks, political parties and the like. In the state of social order, the perception of uncertainty and the rules for coping with it are mutually dependent. In the highway code, for example, this mutual dependence involves social perceptions of traffic risks and the social practices for coping with them. This relationship of mutual dependence guarantees the social acceptance of traffic accidents. The protests against the transport of radioactive wastes in Germany, in contrast, reveal a strong discrepancy between the social perception of the uncertainty in exploiting nuclear power and the way in which society copes with it. In the long-term, this may well compel us to abandon nuclear power completely.

At the same time, the mechanism of the closure of social cause and effect is responsible for the *emergence* of the social (cooperation instead of confrontation). It is accompanied by the exclusion of an environment. According to our construction, this consists of those individuals who act outside of social regulations. Operational closure generates the "antisocial" and the "social" individual; both stand face to face in the relation between system and environment.*

Uncertainty cannot be "absorbed".[10] It is a continuous threat to the established forms of social coexistence. Various types of uncertainty can be distinguished, because they address a completely specific need for regulation within the social community. For example, any society has to regulate the individual-society difference: that which is assigned to the domain of individual responsibility and that which has to be regulated through collective norms; in other words, that which should be regulated through policy and that which can be left to the private initiative of citizens. Alongside such *uncertainties of regulation*, there is also uncertainty of codification: the binding statutory determination of an offence in public and civil law. For example, is the Berlin Love Parade an expression of individual joie de vivre or a threat to public order? Such *uncertainties of codification* characterise the domain regulated by the legal system in modern societies. Before monetary economies developed, there was always uncertainty in an exchange economy about whether one would be offered the object of exchange one wanted when one wanted it. This led to an *uncertainty in supply* that could be reduced to a socially acceptable level only through the medium of money.[11] The uncertainty about knowing whether one is loved, not knowing what makes the world go round or why fortune is (mostly) kinder to others opens up new fields of uncertainty whose regulation is the precondition for an organised coexistence. The topic is *uncertainties in love*, *uncertainties over knowledge* and *uncertainties about beliefs*. People try to save themselves from their doubts about love through marriage, which, in turn, cannot guarantee love. Uncertainty about knowledge of

*Luhmann[9] distinguishes between the psychological and the social system.

the world is reduced by science.* And the only socially acceptable way to shape doubts regarding the justness of the world is to refer to a higher level of justice in another world.

This separation into the various forms of uncertainty has not been performed for analytical reasons alone. The crucial aspect is that the social activities considered suitable for helping to reduce various types of uncertainty clearly differ from each other and are not interchangeable. Uncertainties in supply cannot be regulated through beliefs, nor uncertainties about knowledge through money, nor uncertainties of codification through knowledge. The various forms of uncertainty differ through the social practices used to cope with them. And new forms of uncertainty, in contrast to threat, are perceived as risks only when corresponding techniques and methods for coping with them have emerged in society.** The threat to life and limb of evil spirits can be confronted only when persons are available who are considered to be qualified for this task and who can make us believe that evil spirits can be appeased through sacrifice. Likewise, uncertainties in beliefs regarding how the world goes round could only be tackled as uncertainties of knowledge when, at the birth of modern science in Europe, experimental procedures were developed with the aid of manual techniques already available in other contexts. This made it possible to transform the uncertain knowledge of belief into the certain knowledge of the natural sciences and distinguish questions of belief from questions of knowledge. And when, as in, for example, ancient China, knowledge did not pursue the goal of explaining the world, but understanding it, a natural science comparable with that in the West did not emerge, even though the technology was available, but procedures of contemplation directed towards unity with nature and this form of understanding. These alternative forms of knowledge are now becoming popular in Europe as the risks of technical knowledge are re-evaluated. Acupuncture and high-tech medicine work side by side in peaceful coexistence.

These brief indications must suffice to show that social differentiation is accompanied by the continuous emergence of new forms of uncertainty or with a fundamental change in existing risk landscapes. In completely general terms, a change can be observed in society from administering the present to designing the future. This transition changes functioning systems of regulations and creates new uncertainties. In business, the uncertainty about introducing an existing product to the market profitably is being replaced increasingly by the uncertainty of having to develop this product in the first place. This transition from the economic uncertainty of traditional marketing strategies to an innovation uncertainty in product development is associated with a restructuring of classic organisations. Flat hierarchies, slim-line management and network-like forms of distribution of labour are the characteristics of success in modern companies. Economic development is determined increasingly by knowledge, and, thereby, the generation of knowledge within the economy.*** In science as well, the transition from explaining to constructing the world has produced new forms of knowledge uncertainty that have engendered serious changes in the research landscape. There is even talk about a "new production of knowledge".[14] This new form of knowledge production, tied closely to applied contexts, has led to a move from academic research to different fields of practices. As a result, the self-control within science, mostly institutionalised within academic research with its disciplinary transparency, has become increasingly problematic and is being replaced by forms of state regulation.****

*Although, even then, we can never be certain, and uncertainly assails us again, at the latest, when two experts offer us contradictory opinions.

**On the difference between risk and threat, see Luhmann.[12]

***See Willke:[13] "This knowledge production outside the institutions of the science system that are responsible for it has far-reaching consequences for the science system".

****It is only necessary to consider the efforts to achieve a political regulation of research in the field of modern genetic technology that, in contrast to what its name suggests, is now a part of basic research in modern biology.

INNOVATION NETWORKS

In recent decades, many areas of society have seen the emergence of a new way of coordinating social activities alongside the classic forms of the market and organisation. This is described with the term network. According to Kowol and Krohn:[15] "The key concept of networking found in the scientific research accompanying this new type of phenomenon is hard to specify exactly. In the social sciences, the terms is always used when social interaction contexts are observed that overlap or are in juxtaposition to individual social formations (groups, organisations, system) and have their own form of social integration. Clear demarcations (above all, formal organisation borders) are not necessary for networks. Various persons may be involved at different points in time; likewise, as vibrant networks of relations with changing members, they lack any centre."[16]

"Technological innovations", in particular, are organised increasingly through "innovation networks" these days, because these are, so to say, "model markets for prototypes that develop under carefully negotiated framing conditions. The risk of a lack of market transparency for new products is finally resolved in innovation networks in which a specific constellation of actors work together in tackling the design of new products and the formation of new application contexts, new technologies and their markets in order to further develop them into a competitive form in the case of success."[17]

This reduces the uncertainty over the marketability of innovations considerably, because economic rationality is already integrated into the phase of construction.

Opinions differ in the literature as to why such innovation networks have emerged. However, there is consensus that innovation networks are associated with the characteristic of functional differentiation in modern societies: innovation networks reduce the risk of social differentiation leading to an inability to bridge the gap between each specific system rationality. This argument overemphasizes the "rational" closure of functional subsystems, and overlooks the basic problem confronting every innovation: the production of new knowledge within the context of a planned application. This problem will be addressed in more detail in the following.

In completely general terms, a new idea can only assert itself when, within a certain context, it promises to solve a problem that is perceived as such in this context. A new theory in science remains insignificant as long as its author fails to convince the scientific community that he or she has made a major contribution to solving an important problem. In this sense, the new must be capable of linking up with a social practice in which problems and their solutions are perceived in a specific way.* Technological innovations as well can assert themselves on the market and achieve economic success only when they are able to connect with a social practice of problem definition and problem solution.

One example for the dependence of an innovation on its applied context is the development of the sextant. In navigation, it had always been impossible to fix one's position precisely because no procedure was available to ascertain longitude. This uncertainty in locating one's position often had fatal consequences. Ships broke up on cliffs, failed to reach important stations for taking on fresh water and provisions, or reached them too late. One rather obvious solution to this problem was to measure the interval between two local times: the local time at a location with a known geographical position such as Greenwich and the local time on the ship. The time difference could be used to find the longitude of the ship: three hours later than Greenwich means 45 degrees west of Greenwich. As latitude and ship time had always been ascertained by measuring the heights of the culminations of stars, an astronomic procedure was also sought for the measurement of longitude. The permanently installed ship's clock with the reference time

*One example from science is quantum mechanics. Although it solved an important problem in atomic spectra, it had great difficulty in gaining acceptance because it contravened established habits of thought in physics.[18]

for a place with a known longitude contravened established practice in navigation, and was therefore rejected in favour of the sextant that could be used to ascertain the ship's time and, by measuring intervals between stars, longitude.

As this example shows, innovations cannot be developed in an ivory tower. The sociotechnological environment in which they should be applied plays a decisive role in their construction. The relevant knowledge for an innovation can be obtained only with reference to its application. Michael Gibbons et al.[15] have also pointed to this relation between knowledge production and applied context in his book, "The New Production of Knowledge". He finds that the problem of applying scientific knowledge in non-scientific contexts is solved by an institutional shifting of knowledge production into applied practice. In the present context, we are not interested in how this shift of knowledge production changes the institutional structure of science, but in why this problem in the application of scientific knowledge has suddenly become so critical.

Up to now, scientific knowledge has been assigned a universal validity regardless of where it has been produced, and it has been granted a problem-solving capacity that is independent of the concrete context. It is only since laboratory studies that we have become aware of the fact that a successful transfer of knowledge is tied to the transfer of the context of the knowledge production as well.[19] The scientific laboratory proved to be the classic location for the production of new knowledge. It first has to be set up as a context in order to confirm knowledge produced at another location and to use this as a resource for its own production of new knowledge. If this reconstruction of the context is unsuccessful, "laboratory knowledge" is of little help. It is available to all users only when it is a universal form of theoretical knowledge that has been purified of the context of its origins. Its application in the concrete case then requires a re-contextualisation: the integration of the new applied reference in the general form of the knowledge. The effort involved in this re-conceptualisation is particularly high when complex, non-linear problems have to be solved. In contrast to linear problems, in which small changes in the characteristic parameters always also have small effects, the relation between cause and effect is more complicated in non-linear problems: small changes *may* have large effects. The complexity of non-linearity lies in the indeterminacy that does not *necessitate* a large effect after a small cause but makes it *possible*. Through this *indeterminable* relation between cause and effect knowledge is transformed suddenly into a source of uncertainty.*

This uncertainty determines the role of the context in the production of new knowledge. In linear problem states, in which similar causes have similar effects, the effect of the context on the knowledge can be determined, and the knowledge can be applied to any context. In the formal language of mathematics, this means that a linear problem possesses a general solution with free constants, and every context can be taken into consideration by specifying these free constants. The law of gravity describes, in general, how a body moves under the influence of an attractive force. If we want to know the velocity of free fall of a body at a certain point in time in a concrete case, all we have to do is take the general law, insert the initial conditions and the value for the acceleration due to gravity valid at the selected location, and compute the speed.

This simple relation between regularity and context is no longer given in complex problems. Slight fluctuations in context may have large effects on the solution to a problem. The knowledge that describes a content adequately in one context may prove to be false in a slightly changed context. A general solution to complex problems for all contexts does not exist. Each context has its own specific solution: nothing can be learned from similar situations.

To summarise, with simple problems, it is possible to find solutions that can generalise

*It is important to emphasise that this uncertainty is not a consequence of knowledge gaps. A non-linear regularity describes a certain content exactly; however, in the concrete case, this knowledge does not lead automatically to knowledge of its dynamics.

to all contexts. This is a direct consequence of the definition of simplicity chosen here: linearity. In complex problems, solutions are extremely context-dependent because of non-linearity. The generalisation of the knowledge from comparable situations to the concrete case is loaded with uncertainty because of the indeterminacy of behaviour linked to non-linearity.

Innovations belong to this category of "complex problems". Their complexity is a consequence of the close ties between three problem dimensions: an innovation has to function, connect with a social practice, and succeed economically. The non-linearity of this set of problems becomes apparent immediately when it is considered, for example, that although slight technological modifications may improve economic success, they may well threaten the ability to connect socially. Thus, innovations are always loaded with risk. The non-linearity of the task description prevents the innovation task from being broken down into subsidiary problems and solved at different locations with different competencies. In other words: the technical department develops a prototype, the production department looks for economic production methods and appropriate materials, the marketing department deals with the product launch, and the sales department looks for an appropriate distribution network. If all departments perform optimally, a new product emerges with which money can be made: economic success is guaranteed.

This classic form of coping with innovation tasks within commercial organisations has given way to a new form: innovation networks have emerged and prove to be successful in coping with innovation uncertainty. The rest of this section will cast more light on the reasons for the effectiveness of this new social structure. The previous section has shown that self-organisation in the domain of the social emerges through coping with uncertainty. Through this, it develops not only a specific form of social cooperation for coping with uncertainty but also distinguishes itself from other forms of cooperation (system formation). At the same time, this form is structured so that it will prove successful and reduce uncertainty (structure formation). Hence, innovation networks are forms of social self-organisation that emerge or are applied to cope with innovation uncertainties. The difference between self-organisation and self-regulation - meaning that it is set up externally - is irrelevant here. Furthermore, the establishment of innovation networks can be conceived as a political instrument to control scientific-technological progress. Innovation networks are social tools for solving complex tasks. They are operationally closed in the sense that the perception of innovation uncertainty within the network generates social activities to reduce this uncertainty. There is neither an externally given plan of how to proceed nor a set measure for evaluating the uncertainty that still exists at a given point in time. An innovation is found when all persons participating in the network believe that the product seems to confirm the implicitly reached assumptions on its ability to function, connect socially and be economically profitable. One can never be completely certain about this. One will only really know whether one's assumptions were correct after a successful market launch.

Because of their circular causality, innovation networks are operationally closed social systems that are, nonetheless, open to the resources they require for their work: the various stocks of knowledge for solving the innovation problem. However, the solution cannot be assembled from existing stocks of knowledge. Existing knowledge serves only to define problems and develop search strategies for the production of new knowledge that is the precondition for constructing an innovation. Innovation networks generate new knowledge from old; purposeful meaning from structural information.

The resource of knowledge can be made available to innovation networks in different ways: first, as information on databanks and information systems; second, through the expertise of their members. The latter also guarantees access to implicit forms of knowledge that also take on an important role in the innovation process. Innovations also have to succeed, and this calls for ability in handling the world. Because innovation networks have to integrate different competencies, they cannot be closed in terms of membership. Anybody who has something qualified to say can become a member. This distinguishes innovation networks from classic organisations in which membership is generally regulated contractually. Innovation networks have emerged in order to overcome the limitations of organisations when coping with innovation problems. It is precisely in fields in which the problem is defined, the resources for solving it are known, and the division into sub-problems with a restricted range succeeds that organisations are able to develop their unbeatable capacities. Complex problems, in which even the problem itself does not have a clear definition, produce the organisational failure that is mentioned so frequently.[20] Innovation networks are the only exception here because of their self-organisation. It does not matter whether they are viewed as self-organising in the strictest sense, in other words, as not having been set up externally, or as only self-regulating. In the former, it can be anticipated that operational closure is a consequence of the perception of innovation uncertainties that has its starting point in the network. In the latter, circular closure is organised externally as in the thermostat. The emergence of innovation, just like the formation of structure in natural systems, depends on operational closure alone.

REFERENCES

1. J.H. Turner, *The Stucture of Sociological Theory*, Dorsey, Chicago (1986).
2. G. Kiss, *Einführung in die soziologischen Theorien*, Westdeutscher Verlag GmbH, Opladen (1993).
3. N. Luhmann, Selbstorganisation und Information im politischen System, in: *Jahrbuch für Komplexität in den Natur-, Sozial- und Geisteswissenschaften*, U. Niedersen ed., Dunker & Humblot, Berlin (1991), p. 28, translated.
4. N. Luhmann, Selbstorganisation und Information im politischen System, in: *Jahrbuch für Komplexität in den Natur-, Sozial- und Geisteswissenschaften*, U. Niedersen ed., Dunker & Humblot, Berlin (1991), p. 13, translated.
5. Gierer, Physik der biologischen Gestaltbildung, in: *Selbstorganisation. Die Entstehung von Ordnung in Natur und Gesellschaft*, A. Dress, H. Hendrichs and G. Küppers, ed., Piper, München (1986).
6. H. Haken, *Erfolgsgeheimnisse der Natur: Synergetik: Die Lehre vom Zusammenwirken*, 3th Edition, Dt. Verlags-Anstalt, Stuttgart (1983).
7. G. Küppers, Selbstorganisation: Selektion durch Schließung, in: *Chaos und Ordnung: Formen der Selbstorganisation in Natur und Gesellschaft*, G. Küppers ed., Reclam, Stuttgart (1996).
8. H. von Foerster, *Ovserving Systems*, Intersystems Publications, Seaside, Cal. (1984).
9. N. Luhmann, *Soziale Systeme. Grundriß einer allgemeinen Theorie*, Suhrkamp, Frankfurt am Main (1984).
10. K. P. Japp, *Soziologische Risikotheorie: Funktionale Differenzierung, Politisierung und Reflexion*, Juventa-Verlag, Weinheim (1996).
11. N. Luhmann, *Die Gesellschaft der Gesellschaft*, Suhrkamp, Frankfurt am Main (1997).
12. N. Luhmann, *Ökologische Kommunikation: Kann die moderne Gesellschaft sich auf ökologische Gefährdungen einstellen?*, Westdeutscher Verlag GmbH, Opladen (1986).
13. H. Willke, Das intelligente Unternehmen, in: *Beratergruppe Neuwaldegg*, Service Fachverlag, Wien (1995).
14. M. Gibbons et al., *The New Production of Knowledge: The Dynamics of Science and Research in Contemporary Societies*, Sage, London (1995).
15. U. Kowol and W. Krohn, *Innvoationsnetzwerke*, IWT_Papers 6/94, Bielefeld (1994).
16. U. Kowol and W. Krohn, *Innvoationsnetzwerke*, IWT_Papers 6/94, Bielefeld (1994), p. 13.
17. U. Kowol and W. Krohn, *Innvoationsnetzwerke*, IWT_Papers 6/94, Bielefeld (1994), p. 24.
18. G. Küppers, *Die Nobelpreise in Physik und Chemie 1991 - 1929: Materialien zum Nominierungsprozess*, Report Wissenschaftsforschung 23, Kleine Verlag, Bielefeld (1982).
19. K. Knorr, The organization of Embeddedness: a constructivist approach to micro-macro relations, in: *Varieties of Structural Approach*, J. Smatka ed., in print (1997).
20. U. Mill and H.-J. Weißbach, Vernetzungswirtschaft. Ursachen, Funktionsprinzipien, Funktionsprobleme, in: *ArBYTE: Modernisierungs der Industriesoziologie?*, U. Mill ed., (1992).

PANEL SUMMARY: MULTISTIMULI AND DIFFERENT SENSORY MODALITIES

Panellists: Giorgio Grisanti (*chairperson*)[1] and Silvana Vallerga[2]

[1]Institute of Biocommunication, Dept. of Audiology
University of Palermo
I-90100 Palermo, Italy
[2]Istituto di Cibernetica e Biofisica
Consiglio Nazionale delle Ricerche, Genova and Oristano
and International Marine Centre, Oristano
Italy

INTRODUCTION (*G. Grisanti*)

Our everyday environment in which we live is characterised by stimuli capable of eliciting our actions (or reactions). Generally our actions are influenced by a number of stimuli which are multimodal.

The effect of multimodal stimuli on recall was studied by Kobus, Moses and Bloom.[1] It appears that the simultaneous presentation of redundant stimuli in multiple modalities does support the multiple-resource hypothesis by displaying enhanced recall when information is available from multiple attentional resources.

VERTEBRATE SENSORY SYSTEMS: ONSET OF EMERGENCE AND ATTENTION (*S. Vallerga*)

Life originated from chaos and reacting to mutable stimuli of chaos, life become more interactive with the outside world and more complex. Nuclear membranes enclosing a replicating molecule as RNA, subjected to the effect of a changeable environment, shaped different life forms. From the primeval sensory system of the common ancestor, still detectable in homeobox genes conserved in species as far apart as fruit fly and human being, complex interacting sensory systems evolved to make more intelligent and flexible the relationships with the outside world. Emergence, attention and creativity are the result of increasing complexity of sensory systems and central neural processing. As the universe grows colder and more structured, life grows more interactive and complex, as Asimov's last question states: life is integration in the universe.

Human and Machine Perception 2: Emergence, Attention, and Creativity,
edited by Cantoni *et al.* Kluwer Academic / Plenum Publishers, New York, 1999.

Stimulus detection

The nature of the stimulus dictates the mechanism of detection in sensory systems. Vision and olfaction have quantal mechanisms of detection, due to the quantal nature of photons and odorant molecules. The molecular mechanisms of transduction of the stimulus into a sensory response are amazingly similar for the two sensory systems. The receptor molecules, capturing the sensory stimulus, in both cases belong to the G-protein superfamily. Photon absorption and odorant molecule binding initiate a G-protein cascade of events leading eventually to the sensory response.[2] The energy change involved in the quantal stimulus is in the same order of magnitude. Single photon absorption and single odorant molecule binding provide an energy change of 57 kcal/mol (for λ=500 nm) and 10 kcal/mol, respectively, much larger than thermal noise (0.57 kcal/mol).[2-4]

Hearing differs with respect to vision and olfaction in the nature of the stimulus, and thus in the mechanism of transduction, as well as in the energy involved in the transduction process. Hearing has a classical mechanism of sensory transduction due to the classical nature of acoustic waves. The elementary energy of an acoustic wave at 10kHz is $3x10^{-6}$ cal/mol is several order of magnitude lower than thermal noise.

Sensitivity

The sensitivity (S=dr/dI) is measured, in vision, as pA per photopigment molecule isomerised, in olfaction as pA per odorant molecule bound, in hearing as mV/Pa, unit of sound pressure. The sensitivity in vision and olfaction is so high that a quantal event elicits a sensory response.[2,5] The absorption of a single photon by any one of the $2x10^9$ rhodopsin molecules in photoreceptor cells causes the closure of 1/20 of all channels in the rod outer segment, a spectacular feat of amplification producing a detectable sensory response. The binding of a single odorant molecule "is likely" to produce a quantal response in the range of 0.3-1 pA, a sensory response.[3] In both systems the ultimate sensitivity has been reached, a response to the smallest stimulus. In hearing the signal has to be detected against the thermal noise.

Adaptation

Adaptation is defined as a decrease in receptor sensitivity occurring with a maintained stimulus. The kinetics of the sensory response is affected as well. The application of steady light decrease sensitivity and accelerate kinetics of the visual response; the same effect are induced by a persistent odorant in the olfactory system. In hearing the application of a steady deflection of hair cell bundle shifts the operating point of the transduction machinery.

Multi-neural coding and multiplexing

The visual system is one of the most complex and best studied sensory systems. Some of the mechanisms disentangled in the retina have been found to be common to other sensory systems. In the retina each neurone processes part of the sensory signals and through physical contacts or chemical teleconnections interact with the other components of the neural network. Neighbouring cells of the same neural type work together in a co-operative way, far way cells provide an antagonistic surround for a better contrast discrimination of the stimulus. The retina is organised in a number of functional modules to better code the outside visual scene.

Multiplexed coding is a specific adaptation to low sampling density, as in the retina of lower vertebrates where a few large retinal neurones sample the visual scene, or for low density neurones of the mammalian retina, as the y-cells in cats, that exhibit cross-correlations It is known that multiplexed signals reach the optic nerve, but there is not yet clear evidence that multiplexed signals are decoded at higher centre.[6] Conceivably a multiplexed code was an early evolutionary invention that has been conserved for low-density cells, later supplemented in mammals by point-to-point coding using a large number of small neurones (for a review see S. Archer et al. eds.[2]).

AUDITORY-VISUAL CORRELATIONS AND INTERACTIONS (*G. Grisanti*)

Among all the possible correlations and interactions between stimuli of different origin, the most important ones, from a neurophysiological point of view, are the auditory-visual correlation and interaction. Stein, Wallace and Meredith[7] have reviewed some interesting behavioural as well as neurophysiological evidence pertaining to auditory visual interactions. These studies have demonstrated that, under at least some conditions, stimuli that are subthreshold in one or the other modality alone, can elicit a reaction when they are combined, hence showing some nonlinear interactions. Stein and colleagues have provided neurophysiological evidence that neurones within the deep layers of the feline superior colliculus contain large topographic maps of both visual and auditory space, and that enhancements of response are observed to combined auditory and visual stimulation; this structure may therefore be one substrate for integration of the two modalities.

In addition, it has been shown that inputs from unimodal neurones within polysensory cortical areas are important determinants of integration of response.

It is quite clear that the visual-auditory interaction has a very important role in determining the human behaviour. For example, during a storm, the concurrent stimulation of hearing (thumbs) and seeing (blitz) may induce a faster and prompter reaction of escape.

A certain interest in auditory-visual correlation stems, for instance, from the researches on effects of music in film perception. It is well known that seeing the speaker can help a listener to understand the speech in a noisy environment.

The effects of visual input on speech understanding may be entirely due to the fact that the visual input actually supplies additional information about the identity of the speech signal.

David and Hirshman[8] very recently investigated whether the auditory and visual subsystems which mediate priming of comprehension, can interact. They examined the hypothesis that processing in the visual subsystem during an implict memory test permits the access to memories in the auditory subsystem that were stored concurrently with visual memories. Their results are consistent with the view that there is only limited interaction between the visual and auditory subsystems mediating priming.

SOUND ENHANCES VISUAL PERCEPTION

Sound designers in the film and television industry have long relied on the notion that high quality sound can be used to enhance the audience's visual perception of moving images. This concept has never been formally researched or published.

W. R. Neuman[9] set up an informal experiment in which subjects were required to watch two video sequences. In both sequences, the visual information was identical. However, the first sequence was encoded with low quality audio while the second sequence incorporated high quality sound. The subjects were subsequently asked to compare the two sequences. Nearly every subject stated that the second video sequence was visually superior though the only difference was that it had better implemented sound. Neuman explained how "it turns out that there is an audio effect in the evaluation of picture quality. In fact, the effect of improved audio is as strong on picture quality ratings as it is on audio quality ratings".

The audio-effect is often utilised in the field of TV advertising. For example, each of us daily experiences a higher sound level during advertising TV spots. The audio-effect has significant impact also in the design of virtual environment as demonstrated by some of the most widespread video-games in which sound enhances the user's performances.

The sense of hearing is engaging the human sense of sight and is often neglected in the implementation of a virtual world. Hendrix and Barfield[10] have recently indicated that the integration of spatial audio in a virtual environment enhances a user's sense of presence.

A disparity exists between the exploitation of sound and graphics in interfaces. For whatever reasons, the development of user interfaces has historically been focused more on visual modes than aural. A more definite understanding of how sound affects visual perception would allow virtual environment designers to optimally distribute resources among the audio and visual aspects of a virtual world as a method of maximising presence.

Therefore, it would be useful to document the contribution audio provides to the overall sense of presence.

REFERENCES

1. Kobus, Moses and Bloom (1994).
2. S. Archer, Djamgoz, ER Loew, JC. Partridge and S. Vallerga eds., Adaptive Mechanisms in the Ecology of Vision, Kluwer Academic Press (1998).
3. P. Liman, Pheromone transduction in the vomero-nasal organ, *Current Opinion in Neurobiol.*, 6(4), pp. 440-444 (1996).
4. V. Torre et al., Transduction and adaptation in sensory receptor cells, *J. Neuroscience*, 15(12), pp. 7757-7768 (1995).
5. P. Detwiler and Gray-Keller, The mechanisms of vertebrate light adaptation: speeded recovery versus slowed activation, *Current Opinion in Neurobiol*, 6(4), pp. 440-444 (1996).
6. R. Masland, Processing and encoding of visual information, in *Current Opinion in Neurobiol*, 6(4), pp. 467-474 (1996).
7. Stein, Wallace and Meredith (1995).
8. David and Hirshman (1998).
9. W. R. Neuman (1990).
10. Hendrix and Barfield (1996).

VARIABILITY, AMBIGUITY AND ATTENTION: A PERCEPTION MODEL BASED ON ANALOG INDUCTION

Jean-Sylvain Liénard

LIMSI-CNRS
Orsay, France
e-mail: lienard@limsi.fr

INTRODUCTION

Among the many problems that impede our understanding of perception, three appear to be of primary importance: data variability, selective attention, and learning. Data variability is the hidden part of what is called Categorisation by psychologists and Pattern Recognition by engineers. Selective attention is the process used by cognitive systems to locate or identify some entities considered as relevant, disseminated in a set of comparable entities considered as non relevant in a given situation. Learning is the process by which the perceptual system works out the representations of data at each abstraction level, as well as the transformations of information from one level to the next. In the present chapter we shall examine those three problems and present a perceptual model, which aims at integrating them into a single view.

DATA VARIABILITY IN PATTERN RECOGNITION

Considering the present state of the art in speech and image recognition, variability is the common enemy of pattern recognisers. In speech recognition the contemporary approach consists in collecting very large data sets in order to store, compensate or model all the possible variants (due to speakers, recording conditions, accents, emotions) of the entities of interest (phonemes, words, sentences). The sequential organisation of the speech segments and language symbols explains the good results obtained with Markov models. However, it may be observed that the results are acceptable in those situations where variability is entirely controlled, either by an exhaustive storage of the variants or by pronunciation constraints that the speaker is requested to obey. In computer vision one usually deals with two levels. Low level processing aims at extracting relevant features from the image, while some structural considerations are implemented at the high level. Computing robust features from the image is a difficult, error-prone process, because of the many sources of variability governing the formation of the image. Intensity, disposition and general characteristics of the light sources, reflectance properties of the object, curved

Human and Machine Perception 2: Emergence, Attention, and Creativity,
edited by Cantoni *et al.* Kluwer Academic / Plenum Publishers, New York, 1999.

surfaces, shadowing, 3D orientation and position of the object as well as of the camera, all contribute to the extreme difficulty of the low-level processing.[1] Thus artificial vision is mainly confined to situations where the illuminating conditions are well controlled, the shape of the objects is simple enough to be computationally tractable, the camera(s) and object disposition is known, etc.

The puzzling aspect of variability is that it is a problem for machines, not for humans. In natural communication the speech wave not only conveys words and sentences, but also other kinds of information, related to the identity of the speaker, his linguistic origin, his physical and psychological state, his distance to the listener, his intentions and emotions. Most of this information is coded in the variations of the speech wave structures. It is not to be considered as parasitic or superfluous. On the contrary, it seems that all kinds of information transmitted by the signal have to be decoded simultaneously.[2] Similar considerations hold for vision. Even if the goal is to identify an object, the human viewer unconsciously perceives information about ambient light, distance and orientation of the object, surface properties, which helps him in interpreting the retinal image and, ultimately, in identifying the objet of interest. Our point of view is that, instead of looking for invariance that exist only in artificial situations, one has to simultaneously consider all of the perceptual aspects of the object as well as their interactions.

FROM PATTERN RECOGNITION TO PATTERN PROCESSING

Thus we have to question the very basis of the Pattern Recognition paradigm, which is called Categorisation in psychology. Recognising or categorising a pattern implies that each category is made of several patterns (or *exemplars*), that are supposed to be equivalent to each other. Any exemplar of the category can be replaced by a single pattern called *prototype.* It is a "many to one" process. It apparently yields a drastic reduction of redundancy, but the price to pay is variability, because each pattern actually differs from each other. If the differences are random or irrelevant the only consequence is some uncertainty on the class boundaries, which can be treated by statistical techniques. But if the differences between exemplars reflect some relevant information coded in the data, then considering the exemplars as equivalent may produce some unrecoverable classification errors. This is what happens in most pattern recognition systems.

To deal with this problem Pattern Recognition could be replaced by Pattern Processing or, equivalently, Categorisation by Multi-Categorisation. The essence of the Pattern Processing paradigm is to associate each object with its *complete description* in terms of abstract qualities. Instead of simply classifying a pattern into a single set of categories, we propose to associate to it several sets of categories. Each of those sets is an *attribute* or *high-level descriptor*, that take several values. Each set of values (or *description*) is sufficient to define the perceptual information contained in the pattern. In other terms, if two patterns are considered by subjects as perceptually equivalent, then they must have the same high-level description. Conversely, if they differ at the high level, it means that they are perceived as different from each other. Thus Pattern Processing is a "many to several" process. It transforms a low-level pattern, close to the physical stimulus, into a high-level description, which can be viewed as another, more abstract pattern. It aims at recognising not only the identity of the object, but also all of its properties that have some perceptual relevance for the human observer of the same signal or scene. In Pattern Processing not all of the variability is eliminated: there remains some *residual variability* which represents the variability considered as non-relevant for the human observer. The difference between Pattern Recognition and Pattern Processing is illustrated in Figure 1.

Pattern Processing is not opposed to Pattern Recognition. In fact, Pattern Processing can be reduced to Pattern Recognition by specifying only one descriptor at the high level. The discrete values of this descriptor represent the Pattern Recognition categories.

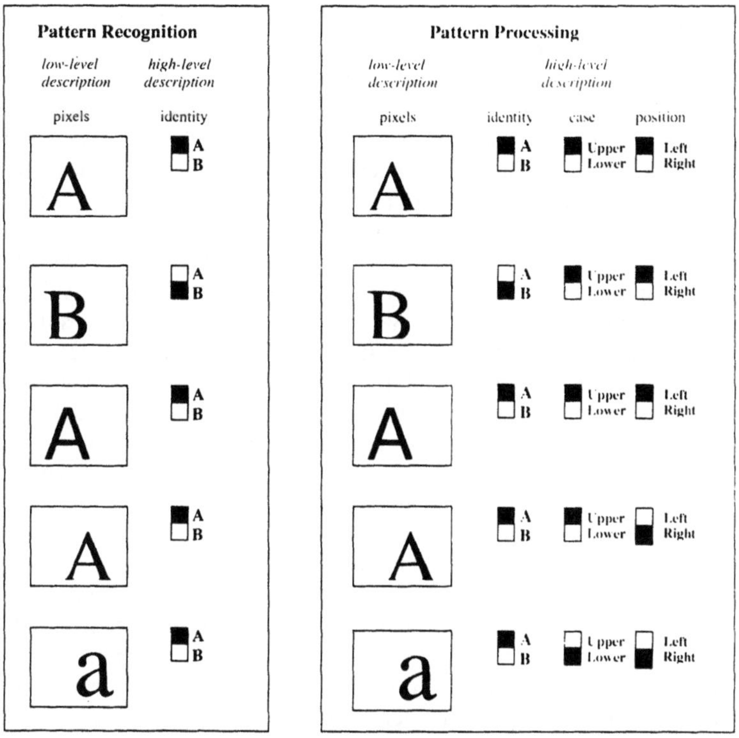

Figure 1. Pattern Recognition (left) versus Pattern Processing (right). Pattern Recognition implies a large variability of data, as can be seen by the diverse low-level representations of the character "A". By specifying several high-level descriptors instead of a given set of labels, Pattern Processing reduces variability. The low-level patterns which have the same high-level description are considered as equivalent at any level. Their differences at the lower level constitute the residual variability.

Conversely, Pattern Processing can be viewed as a kind of Pattern Recognition which classifies the low-level information into a number of categories equal to the product of the numbers of discrete values that the high-level descriptors could take. However, the specificity of Pattern Processing is twofold. Firstly the high-level descriptors reflect the perceptually relevant dimensions of the low-level information. Secondly it poses that at each level the same *complete* perceptual information is represented, with the high-level descriptors more decorrelated than the low-level ones.

The term of "pattern" has to be specified, as it is used here in a wider sense than usual. At the low level, a pattern is the projection of an object or structure of the physical world on a set of sensory captors. It is made of elements (such as pixels or auditory quanta) and topological relations in the projection space (such as vicinity, distance, simultaneity or continuity). It has a functional identity, i.e. it is interpreted as the same entity in different situations or environments. At the high level it may be represented by a single label reflecting this identity and some information concerning its relations with the neighbouring labels in a given conceptual space (contextual information). At intermediary levels it is represented by sets of indices relevant in the definition of a given object.

Representations at the low and high levels both refer to the same object. They are made of two distinct types of information, which specify on one hand some identity features, and on the other hand some localisation features. Let us define any representation level as i) a set of elements (or variables) and ii) a set of rules that define the manner in which the elements are ordered at this particular level. The variables are what we called above identity descriptors; they may, or may not, be instantiated according to what happens at the neighbouring levels. The ordering rules define the proximity of the variables. In some cases these rules may appear as a metric. This is the case at the low levels, where the topology of the outside world is more or less reproduced (retinotopy, tonotopy, stimulus intensity, time dimension). At higher levels some new ordering rules may appear, that reflect the functional properties of the objets. In this case the proximity of two variables may be expressed in a non-metrical form such as the shortest path between two elements of a graph. More generally, as proposed by Gärdenfors,[3] any item at any level has to be located in some conceptual space.

Thus, whatever the level considered, there are descriptors specifying level-specific elements, and descriptors specifying the way in which these elements must be organised if they are to represent objects of the world. We call *pattern* the set of values taken by all the descriptors of a given level in response to the presentation of an object known to the perceptual system and validated at the high level. In this perspective, Pattern Processing is the process which computes all the descriptors values of a given level, from the knowledge of all the descriptors values of the adjacent levels. It is more general than Pattern Recognition, which aims at reducing from level to level the number of descriptors, until there remains at the high level only the identity descriptor of the object.

The above definition of a pattern does not in itself imply the existence of localisation descriptors: any given set of values taken by a set of independent variables can be considered as a pattern (in this case made of identification descriptors only), provided that it can be related to a given object of the world. However, the identity of an object can be stated only if this object can be declared as the same entity in different contexts (at least two). Thus, patterns usually comprise localisation descriptors (to ensure the identity of objects in different contexts) as well as identification descriptors (to distinguish between different objets).

VARIABILITY, AMBIGUITY AND SELECTIVE ATTENTION

Let us now consider a two-level perception model. At the low level we have a representation very close to the signal itself, such as pixels in the case of vision or time-frequency elements in the case of audition. At the high level we have abstract properties of the signal, which usually consist of several descriptors specifying the properties and localisation of the object in the input space. For instance a set of low-level pixels whose meaning (for the human observer) is a red square located at the top of the input screen is described at the high level by an identity descriptor (that could takes several values such as triangle, square, circle), a colour descriptor (taking values such as green, red, blue, black) and localisation descriptors taking values such as top/bottom, centre/left/right. Bottom-up functioning of the model consists in activating the appropriate pixels and obtaining the complete description of the object. Top-down functioning starts from a high-level specification and provides a low-level representation in terms of pixels. In usual situations the information provided at both levels is partial, and the aim of the model is to complete both descriptions.

This model illustrates the duality that exists between variability and ambiguity. Variability occurs when several input patterns are associated with the same high-level description. Similarly, ambiguity occurs when a given low-level pattern can be associated with several high-level descriptions. There is a deep symmetry between both notions, with the difference that low-level patterns usually reflect data coming from the external world, while the values taken by the high-level descriptors can be set or modified by the cognitive system to which the perceptive system is connected.

Instead of considering only two levels one can consider a hierarchy of representational levels. It is a necessity in speech and voice processing, because there is some evidence that the perceptual information is structured at different levels such as auditory indices, features, phonemes, words, concepts, all differing in their time resolution.[4] Thus perception can be viewed as two simultaneous information flows moving in opposite directions, bottom-up (data-directed flow) and top-down (concept-directed flow). At each level the appropriate descriptors are represented by units whose activation values form a constantly evolving pattern. Each unit tends to combine the different pieces of information coming from its intra- and inter-level neighbours in order to elaborate the right values and to send them back to its neighbours. In this process information tends to complete at each level, until it changes under the pressure of one or the other flow.

In neurobiological systems evidence has been found of the existence of different pathways from the primary projection area V1 to the higher vision areas, specialised in identification ("what") and localisation ("where"). There is also some evidence that most bottom-up connections have their top-down counterparts. Psychologists[5] have shown that selective attention manifests itself virtually (covert attention) before producing a shift in the eyes orientation (overt attention). In covert (or static) attention two different mechanisms (parallel, or "pop-out", and serial search) seem to be activated, according to the way the element to be identified (target) differs from other elements (distractors) present in the visual scene.

In everyday life selective attention is crucial. The visual field is permanently filled out with various stimuli that must be selected according to criteria requested by the cognitive system, except when for some reason they impose themselves to the attention, for instance through a rapid movement. The perception model described above provides a framework adequate to study the phenomena of selective attention. In the presence of a low-level scene comprising several objects, covert attention is simulated by fixing the values of some high-level descriptors, which partly define the properties of the particular object one is interested in. Running the system yields simultaneous activation at the low level (the pixels corresponding to the object of interest keep activated, the other ones are deactivated) and at the high level as well (the values of the descriptors which where unspecified in the first phase get fixed). Simulation of overt attention (eye saccades) may also be achieved with two adjunctions: a low-level information representing the eye orientation with respect to the whole visual space, and a mechanism able to move the eye optical axis toward the area of interest.

ANALOG INDUCTION

The third question raised in the introduction deals with learning. This term has to be precised in the context of our model. It may apply to the set of descriptors represented at each structural level. In some cases one can guess the nature of the descriptors. For instance neurobiological studies have evidenced some cells in the visual system that respond to specific stimuli, in various cortical areas. In the speech perception domain, linguistic studies have assessed the relevance of the notion of phonetic unit as well as the notion of word. This knowledge gives some clues to enumerate plausible sets of descriptors at various levels. Investigating how the brain has acquired those descriptors is a first learning problem, which concerns biology as well as developmental studies.

We shall not consider this problem in the present paper. Instead, we shall suppose that the set of descriptors is given, at least at some of the structural levels taken into consideration. For the sake of simplicity we shall consider only two levels. Thus the learning problem we shall address here is to work out a process by which the low and high-level descriptions could be associated, given a set of examples.

In the Pattern Processing paradigm the availability of a complete description at each level allows the learning of transformations of the low-level representations, which

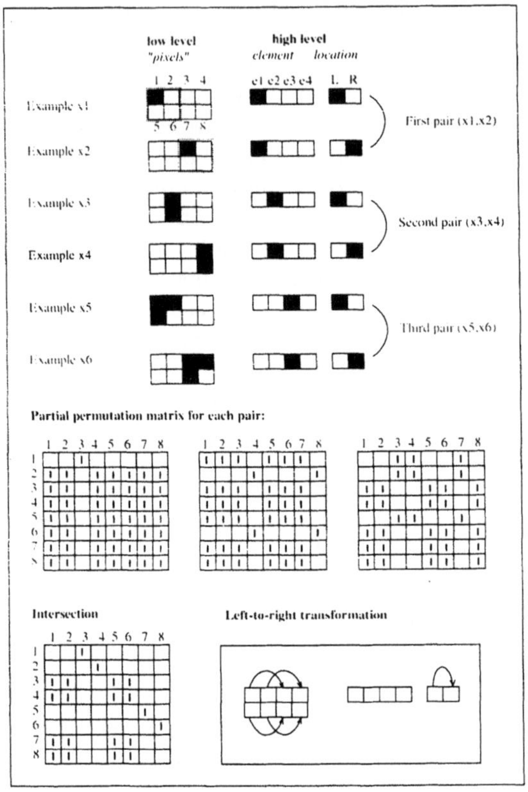

Figure 2. Principle of Analog Induction. Here the goal is to learn from paired examples the partial permutation corresponding to the Left-to-Right transformation of the low-level data.

correspond to changes of values of the high-level descriptors. This process called Analog Induction[6] is illustrated in Figure 2 with schematic data.

The system input is a binary vector divided into several fields. Each field corresponds to a descriptor: in Figure 2 there are one low-level descriptor made of eight binary variables (that we will temporarily call "pixels"), and two high-level descriptors, one indicating the identity of the pattern (one out of four), and the other one indicating its position (Left or Right). The output is another vector of the same format, which in the test phase represents the system answer after it has scanned its memory and retrieved the closest vector. This associative and reactive functioning is of the same sort as the one described by Hintzman[7] in his memory model Minerva2. The learning phase consists of storing a set of training vectors. The induction phase consists of augmenting the memory with vectors representing cases not encountered during the initial storing. The goal is to induce the Left→Right transformation of the low-level pattern.

Among the training data one selects a pair of vectors representing the same element (same value of the identity descriptor) but differing by the position. To this pair one associates a binary square matrix in which the value 1 indicates that an active (resp. inactive) pixel of the input may become an active (resp. inactive) pixel of the output. Another pair of examples selected in the same way (other element but same change of position) yields another matrix, and so on for all the appropriate pairs that can be found in the training examples. The intersection of all these matrices gives the matrix associated to the transformation, in which some lines have a single non-zero value. In certain conditions

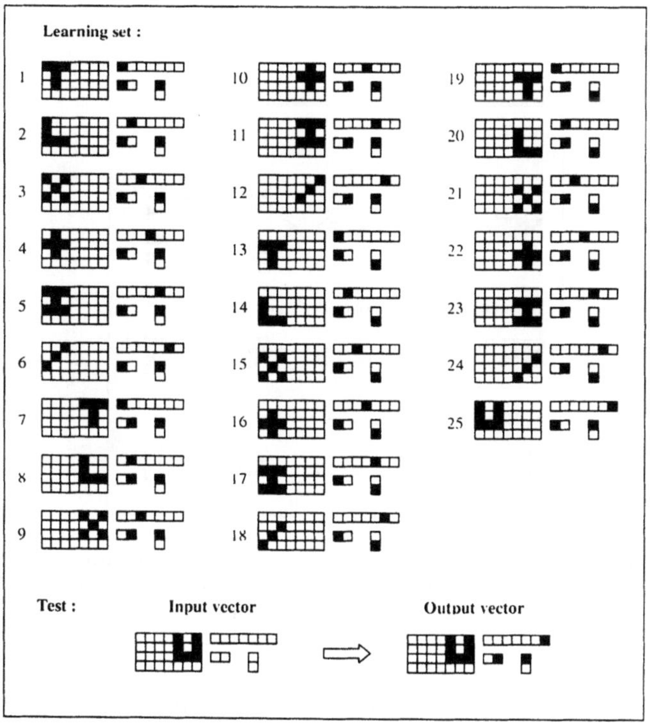

Figure 3. Analog induction with two localisation descriptors. Top: Learning set (25 examples). Bottom: recognition of a pattern (item 7, U shaped) presented in a new position (right).

this defines a partial permutation of the low-level pixels. Then this partial permutation representing the Left-to-Right shift can be applied to any other example encountered only in the Left position during training, and the resulting vector can be stored in memory. Thus the system is able to recognise an example it has never encountered in the Right position during training.

This process is inductive because it works out general relations from particular examples. It is analog because it repeatedly uses the basic scheme of analogy: "A is to B what C is to D". It is well known that analog reasoning critically depends on the property used to match the examples.[8] This usually makes it difficult to maintain the consistency of the results. But here analogy is applied several times to the set of examples and the matching property is explicitly defined by the high-level descriptors. For a given descriptor each matching between two examples is poorly determined, but considering a sufficient number of pairs yields a permutation (or mapping) which has a general value. In other terms this inductive system uses a beam of matchings between example pairs (typically one set of matchings per pair of values of each descriptor), which is made possible by the existence of a complete description at both abstraction levels.

The same process can be applied to problems involving several position descriptors, with the same result, i.e. a pattern solely known by the system in a given position can be recognised when it is presented in any new position (see Fig. 3).

Figure 4 illustrates the desambiguation of an incomplete pattern (intersection of a "+" pattern in the upper part of the low-level field, and of a "T" pattern in the lower part), which could be equally recognised as any one of both elements. If any complementary

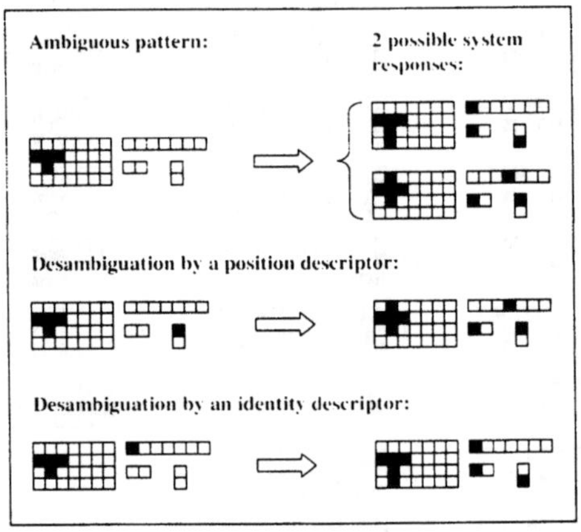

Figure 4. System response to an ambiguous pattern: the incomplete pattern presented at the low level may be interpreted as item 1 (T) in the bottom position, or as item 4 (+) in the top position. The ambiguity is removed by specifying either the position or the identity at the high level.

information is provided, dealing either with the pattern identity or with its approximate location, the ambiguity is removed, the correct pattern is recognised and correctly located.

Conversely, if the low-level field comprises several patterns, even superimposed, giving at the high level some information on the pattern of interest allows the selection of this pattern only (see Fig. 5). The output vector appears then free of the unattended low-

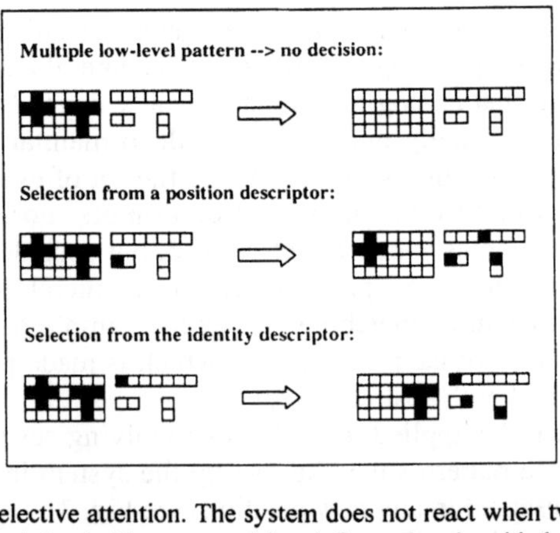

Figure 5. Illustration of the selective attention. The system does not react when two equally probable patterns are presented at the low level (top). If some position information is added at the low level (here: the Left/Right descriptor has the Left value) the corresponding pattern is identified and the position is completed; the unattended low-level pattern is eliminated (middle). If the identity of the expected pattern is given at the high level (here: a T) the corresponding pattern is selected and its high-level description is completed (bottom).

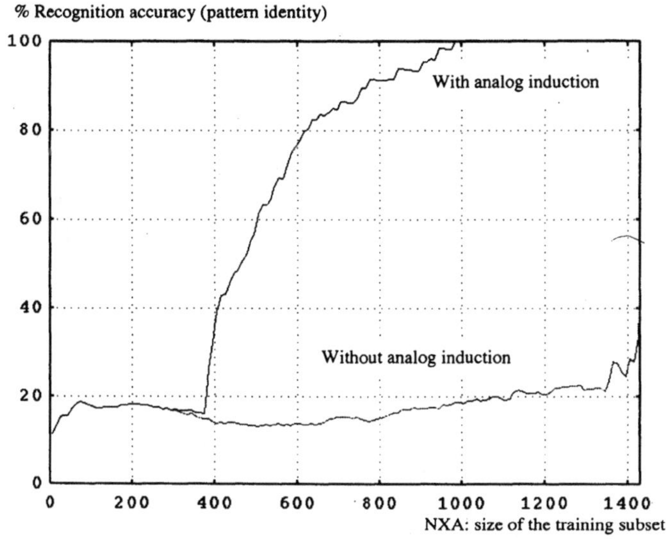

Figure 6. Recognition score versus learning set size, in the case of a database comprising 1440 examples randomly ordered. Without analog induction the recognition accuracy is low, because each pattern appears only once in any position; thus, even when the identity of the pattern seems to be correctly recognised, its position would be wrong. The analog induction process constructs reference patterns from the training subset of size NXA, by applying the learned permutations.

level patterns, and the high-level description gets completed. This is an illustration of the selective attention problem, as explained above.

The Analog Induction process has been tested on a larger set of examples. In one experiment, ten elements were defined at the low level in a 5x5 frame. Each one was presented with 4 degrees of rotation in each of the 36 possible positions in a 10x10 field. The database built in that way comprised 1440 vectors of length 150 binary variables. The vectors were placed in random order. The NXA first vectors were considered as the training set, the rest was considered as the test set and each of the test vectors was reduced to its low-level part (i.e. all high-level information was wiped out). For each value of NXA a recognition score was computed on the test set (taking into account only the value of the identity descriptor), with and without Analog Induction. Figure 6 shows the results. Without Analog Induction the recognition score oscillated between 20 and 30%, due to some regularities in the example patterns which augmented the random identification probability. With Analog Induction the recognition score started abruptly at about NXA=370 and reached 100% around NXA=1000, demonstrating that Analog Induction had implicitly discovered the permutations used to build the data. The abrupt onset of the phenomenon was due to the cumulative aspect of the induction process.

It must be mentioned that the apparent simplicity of the process is only due to the fact that the examples are visually presented as images for the human reader. Actually the system does not use any topology. The term "pixel" is not rigorous to designate the low-level binary variables, because the system does not know that a given low-level binary variable is the neighbour of another one, or lies at a given distance. This means that the process is able to discover some rules in non-metric data, such as the representations that may occur at the higher levels of perception.

Other experiments have been conducted, in order to test the behaviour of the Analog Induction process when dealing with more realistic data. One particular experiment dealt with a systematic variability introduced in the data, without specifying its cause at the high level. Any given low-level element (out of a set of 10) was used as a seed to work out 8

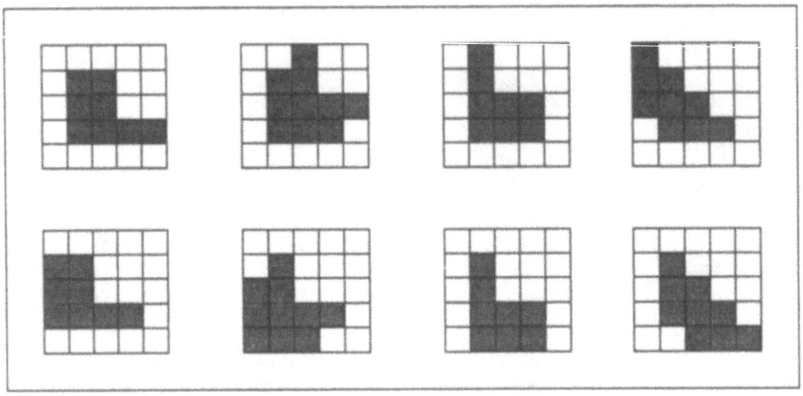

Figure 7. Construction of variants. The character L is never encountered as such in the learning set of examples. Instead, it is implicitly defined by a set of variants built from 8 shifts and unions, which appear randomly in each of the 3x3 possible positions of the low-level grid.

variants, made by a one "pixel" shift and the union of the shifted item with the seed (see Fig. 7).

Thus the information contained in the elements was altered in such a way that they kept some gross similarity but were objectively different when they appeared in different positions. As a consequence it was much harder for the induction process to match the examples and to compute the appropriate transformations from the matrices. It became necessary to slightly change the matrix processing in order to extract the right permutations. The results indicated that Analog Induction could still discover most of the implicit structures of the database, as illustrated by Figure 8. In another experiment, results obtained by using training examples corrupted by noise proved that the inductive process

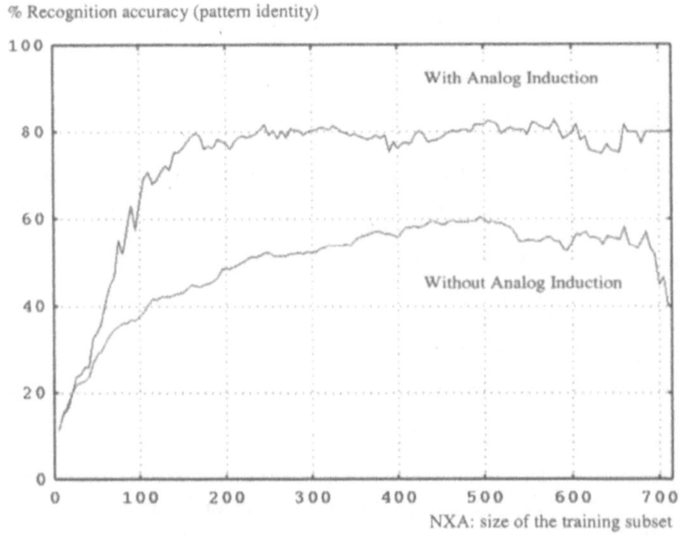

Figure 8. Recognition rate versus learning set size, in the case of 8 variants per pattern. Despite the distortions of the initial L pattern, the induction process has been able to discover most of the permutations used to build the database.

was robust: introducing some noise (pixel inversion) in the low-level data simply increased the size of the training set required to obtain a given recognition rate.

The computational complexity of the inductive process, in the case of a single position descriptor which can take two values, is related to the computation and use of the permutation matrix. It is proportional to the squared number of examples, and to the squared number of low-level variables ("pixels"). When several independent position descriptors are used, the complexity is multiplied by the number of values that each descriptor can take. Thus, although it is not exponential, the cost of the induction phase is high enough to prevent any application to real-size images in the straightforward form described above.

However, some remarks have to be made on this point. Firstly, for a given series of images (representing a given "world", i.e. a set of objects that may appear in different locations or with different distortions), the learning process is made once for all. It is not necessary to learn all the possible cases: once the permutations are learned from a subset of the examples, they stay valid for any new object, which thus can be learned from its presentation in only one position. Secondly, the system described above does not use the two-dimensional topology of the low-level image, nor the topology of the two or three-dimensional world which is implicitly used in the high-level description of the object. Most of the complexity is due to the fact that the system has to discover from the examples some information equivalent to this knowledge, which could be provided by the operator. Thirdly the system is an illustration of a two-level process: it does not use any intermediary representation such as the ones that have been evidenced in vision (orientation detectors, detectors of elementary shapes) as well as in audition and speech perception (acoustic indices, features, phonemes, syllables, words). Work in progress aims at implementing intermediary levels and topologies in order to improve the performance and decrease the complexity, while keeping the properties of analog induction and selective attention.

CONCLUSION

Considerations on data variability in pattern recognition have led us to propose a new paradigm called Pattern Processing, in which any low-level pattern is associated to its complete description at the high level. The high-level pattern comprises two kinds of descriptors, representing respectively identity and localisation information. Perception is then viewed as a hierarchical change of representation with constant perceptive content. It involves two interacting information flows respectively driven by the data and by the concepts. Variability is reduced to the set of variations which are not perceptually relevant. In this view, data variability and high level ambiguity play a symmetrical part. Selective attention appears naturally in this process: it is the consequence of the partial specification of the high-level information. A technique called Analog Induction is presented in order to compute the low-level mappings corresponding to the changes of the high-level localisation descriptors. As a consequence, a system using Analog Induction can implicitly discover some of the structures of the data, and use them to recognise data presented at the low level in new locations. The learning process is robust to some distortions, as well as to the presence of noise.

ACKNOWLEDGEMENTS

The author is grateful to Nassima Boudouani and Youssef Machrouh, for their participation in this work during their DEA stays at LIMSI in 1997 and 1998, as well as to his colleagues, Dr Joseph Mariani, Dr Angel Osorio and Prof. Bertrand Zavidovique, for their support, cooperation and encouragement.

REFERENCES

1. S. Ullman, The visual analysis of shape and form, in *The Cognitive Neurosciences*, M.S. Gazzaniga ed., Bradford books, MIT Press, Cambridge (1995).
2. J.S. Liénard, From speech variability to Pattern Processing: a non-reductive view of speech processing, in *Levels in Speech Communication, Relations and Interactions*, C. Sorin et al. eds, Elsevier, Amsterdam (1995).
3. P. Gärdenfors, Induction and the evolution of conceptual spaces, in *Charles Peirce and the Philosophy of Science: Papers from the Harvard Sesquicentennial Conference*, E.C. Moore ed., University of Alabama Press, Tuscaloosa, pp.72-88, (1993).
4. J.S. Liénard, Perception of speech and voice: beyond pattern recognition, in *Proceedings of the IIASS Vietri Summer School "Neural Nets E.R. Caianello" (1998)*, Springer Verlag ed., (to be published).

5 M.I. Posner, Attention in cognitive neuroscience: an overview, in *The cognitive Neurosciences*, M.S. Gazzaniga ed., Bradford books, MIT Press, Cambridge (1995).

6. J.S. Liénard, Apprentissage de transformations point à point par induction analogique, Notes et Documents LIMSI, 93-07(1993).
7. D.L.Hintzman, Schema abstraction in a multiple-trace memory model, Psychological Review, 93, 4 (1986).
8. D.R.Hofstader and M. Mitchell, An overview of the Copycat project, CRCC Technical Report, 51 (1991).

ALERTING MECHANISMS IN ATTENTIVE VISION SYSTEMS

Giovanni Garibotto

R&D Department
Elsag S.p.A.
Via Puccini,2, Genova - Italy
e-mail:giovanni.garibotto@elsag.it

ABSTRACT

This chapter refers some examples of industrial applications of computer vision where the principle of attentive vision is normally exploited. Actually such mechanism is more common than expected in ordinary situations. This is due to very practical reasons, namely to increase speed in the computation and to improve robustness in the processing to obtain more stable and reliable results.

The biological model of attentive vision is definitely an important reference scheme to follow, being the most effective sensorial system, and many lessons can be learned from it. There are at least two different ways to implement an attentive vision system, using either a bottom-up or a top-down approach. In the first case the low-level features extracted from the scene are used themselves to orient the system attention to a particular area of interest. In a top-down approach the visual recognition task is driven by a specific goal and is constrained within a set of processing rules so that many possible visual routines are intentionally not explored. The examples referred in this chapter range from address interpretation in mail sorting systems, car-plate location and reading in traffic control, up to landmark identification in autonomous navigation systems. The current experience in most industrial application problems proves that top-down schemes are more effective, being able to allow a more accurate control of the system and get rid of possible side effects.

ATTENTIVE VISION

In the last twenty years there has been a lot of development of research in the field of human perception, with great progress in the understanding of the overall process at the different levels. In particular the visual mechanisms at the level of the retina and the transmission of the elementary information at the higher levels has been both demonstrated by the neurobiologist analysis[1] and supported by the mathematical model of computational vision.[2] It is also clear that visual perception is a very complex process which is not only limited to the analysis and processing of the perceived images but it is a combination of different sensorial effects. Moreover, the recent models of purposive vision are clearly supporting human visual perception as a process highly conditioned by expectations and motivations.

In the life of any biological system it is not common to perform a generic passive visual exploration of a scene, without any goal in mind. Actually, the whole perception task, from the selection of the low level processing stage up to the decision and feedback control (eye movement, other sensor activation), is driven by the scope of action (reading a label, looking for something, grasping, etc). Visual attention may be attracted by different objects and details, when the goal is changed, as it is well known from classical experiments,[3] and this happen every day in our ordinary life (looking for a traffic sign when driving, face detection and recognition, etc.).

Computer vision researchers have devoted a great attention to the understanding of biological vision. The scope of such research is at least twofold: supporting the goal of science understanding (biological and psychological research) as well as finding some hints for industrial applications.

Attentive vision in particular has attracted a lot of efforts and contributions in the last few years. There is a wide list of publications on this subject and many web sites refer interesting advanced results (i.e. http://www.klab.caltech.edu/~itti/attention/).

Two main approaches have been developed. The first one is based on the assumption that attentive features should be found automatically from the scene without any particular external constraining mechanism. The second is a model based approach where the objective of the vision recognition task is well defined in advance, and may be more or less accurately modelled (as a 2D or 3D shape).

Bottom-up Saliency Based Vision

The challenge of the unconstrained bottom-up, task-independent approach is the ability to detect saliency points and features that should be automatically distinguished from the background and attract visual attention. Some results are referred in http://www.klab.caltech.edu/~itti/attention/,[4] with a variety of reference applications. The developed model is in accordance with the known anatomy and physiology of the visual system of the macaque monkey and is divided in two steps. A first pre-attentive extraction of visual features (using more than 50 maps in parallel) is followed by a slow and sequential focal attention shifting mechanism. A winner-take-all neural network is used to select the most conspicuous image locations. From the early generic low-level feature extraction process, the training of such neural network should allow a specialisation of the approach to the individual problems at hand.

An interesting example is referred by http://www.klab.caltech.edu/~itti/attention/[4] for the automatic detection of traffic sign labels, with excellent results. They make use of all detectable image feature cues (as colour, edges, contrast, etc.). The basic assumption is that traffic signs are intentionally designed to be easily detected by human drivers in a traffic environment. Then, it should be possible to develop computer vision algorithm to highlight these features and drive the visual attention to such specific regions of interest.

Top-down Model Based Vision

Actually the poor use of bottom-up schemes in the solution of real industrial problems can be explained by the extremely high computational costs (50 saliency maps to be built) and the overall lack of control of the whole detection process. It is worth to remind that industrial vision systems require primarily:

- high reliability and robustness of processing, with minimisation of false negatives;
- synchronisation properties of the sensory events;

• reduction of the amount of information to be processed to save computation and concentrate on really useful and significant features;

• improvement of computational speed to achieve closed loop control.

The pragmatic attitude of any industrial vision system is to solve problems, rather than working with a sound perceptual model behind. That's why the use of all available a priori knowledge is always recommended and often the simplification of the problem is closely matched with the achievement of a robust and reliable result. Currently, multiresolution processing is widely used and closely related to attentive vision schemes, since most of the low-level saliency points can be achieved from low-resolution pictures. Further refinements at higher resolution permit a careful selection of appropriate candidates and removal of outliers.

In the following sections a series of industrial vision examples are referred, trying to highlight the practical use of attentive vision models.

MAIN FUNCTIONS INVOLVING COMPUTER VISION IN MAIL PROCESSING

Mail sorting and postal automation has always represented an important area of application for Image Processing and Pattern Recognition techniques. The most challenging task, performed by vision technology in postal automation, is address reading, including cursive handwriting, flats and parcel image processing and recognition.

Flat handling is definitely a more complex task than letters and cards processing. There are different categories of flats to be handled. One class includes large A4 size envelopes with more or less additional information printed on it (sender and destination address, advertisement messages, stamps and mail class service information). Another class consists of journals, newspapers and catalogues with or without transparent plastic wrapping. Moreover in the flat category are often included also small parcels with a maximum thickness of about 40 mm.

Only very few flat sorting machines are in operation, often without automatic reading capability, but the traffic for this kind of mail is constantly growing, as well as the need for automatic encoding.

One of the main critical functions of address reading is the ability to correctly locate the address block. Quite often the destination address is written on a small label floating under the plastic envelop, which means that it has to be found in any position and with any orientation. Otherwise, the destination address has to be located in a complex scene, full of text and graphics, as newspapers or magazines, where the statistical properties of the address block are very close to the background of the mail item.

Figures 1 and 2 refer a typical example of the problem. The original image (see Fig. 1a) shows a lot of text, with different font-size. The main cues which can be used by a

Figure 1. a) Original grey level image of a flat mail piece; b) connected components detected (blobs).

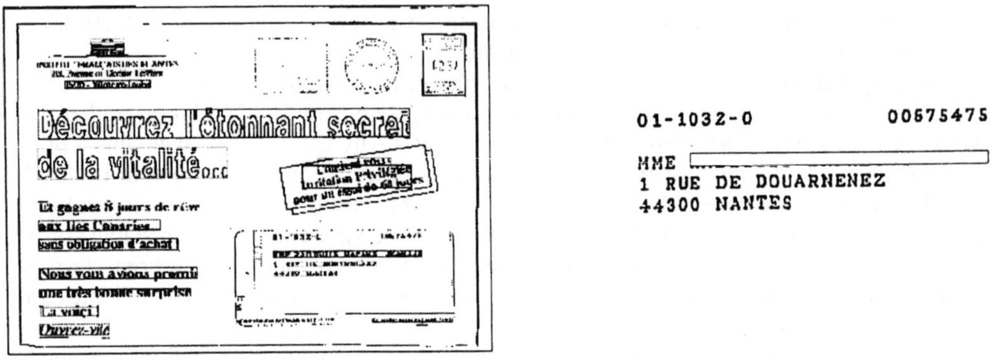

Figure 2. a) Lines detected in the flat image; b) address field detected.

human operator are both the meaning of the text (but this require reading) and some statistical properties (the destination address block is most likely placed within the bottom-right area).

Actually, a detection task which may appear quite easy for the human being, when reading the flat mail at the appropriate resolution, turns out much more uncertain when performed on a low-resolution version (as the blob filtered version of Figure 2). And this is the kind of information which can be used by the automatic detection process, which is supposed to operate at low resolution without any possibility of text reading (unless shifting from low to high resolution in selected ROI fields). Actually the general process of ABL (Address Block Location) can be summarised as follows:

- *multiresolution region of interest analysis*: after pre-processing, to achieve image enhancement and noise reduction, low-resolution processing is performed in order to find regular clusters of patterns (which may be classified as text). Local measures of density and frequency, run-length and blob analysis on binarised images are common tools used at this stage of the process;
- *segmentation of block candidates*: geometrical constraints are commonly used to segment and isolate some blocks and rank them on the basis of similarity measures with respect to the expected address prototype models. The number of detected text lines (horizontal or vertical), their justification (left, centre or right), the size and shape of the block are used as discriminating features;
- *context analysis*: topological constraints as well as heuristic criteria are used to classify the detected blocks and decide on their arrangement onto the mail piece. It is worthwhile to remind that this stage of the process is usually carried out at lower resolution where text characters cannot be recognised and interpreted.

Many research results on this subject, as well on the other topics of mail process automation can be found on the proceedings of the USPS Conference.[5] In the previously referred ABL approach, attentive vision is exploited at the multiresolution stage, to find relevant regions of interest where further detailed investigations and analysis are carried out. Similar processing schemes are used to look for other relevant features on the mail piece like stamps, postage code (airmail, express, etc.), as well as printable areas. Another interesting web site, with relevant results of mail processing and reading, can be found at the CEDAR web page (www.cedar.buffalo.edu/HWAI/).

CAR-PLATE READING SYSTEM IN TRAFFIC CONTROL

The problem of car-plate reading is a classical example of attentive vision. The early systems installed for highway control since the beginning of the 90's, made use of external

Figure 3. Car plate detection and recognition: a) parking access control; b) free-lane control (real-time video-field processing).

synchronisation signals, in order to detect the transit of the car (beginning or end), and optimise the image capture process. Most recent solutions are based on a self-synchronised scheme, where the car transit is detected from the video sequence analysis, without any external control. As such, the vision process can be summarised as follows:

- background video sequence analysis to detect the appearance of a car in the field of views (alerting mechanism) and region of interest as candidates of car-plate shapes;
- investigation of the individual regions of interest to check whether they are really a feasible car-plate field;
- tracking of the recognised car-plate field to gain confidence (time integration).

There are some correspondences with the human visual recognition process, although the practical reasons for this attentive vision process are much more concrete than looking to mimic the biological human visual system. Actually the hierarchy of the computer vision process is mainly due to the need to simplify computation and improve the speed of analysis, as well as to reduce the amount of data to be stored (video sequence) (see Fig. 3).

To speed-up the background video processing, only a subset of image lines are considered to perform an early detection of region candidates, based on a local frequency signature analysis.

VISION BASED NAVIGATION CONTROL

There have been many recent research experiences of vision based navigation control, but just a few industrial applications to AGV's even if this is commonly considered one of the most promising solutions to achieve the necessary flexibility and performance. Most concrete results have been achieved in the automotive industry as a support to the driver, to provide information about the distance from the nearest forward cars and to perform autonomous tracking and car following in traffic congestion.[6,7]

Quite recently, a fully passive vision approach has been proposed to allow self-positioning and autonomous navigation of a forklift carrier, named Robolift.[8] In this case computer vision is used to recognise artificial geometrical landmarks, placed along the navigation pathway and correct the relative odometer estimates.

The landmarks are H-shaped signals (as shown in Fig. 5) stuck onto the floor along the pathway. They are quite easy to install (they can be painted on the floor) and are suitable to be used in a cluttered, quickly changing environment, such as a warehouse. Another advantage of the vision approach is that the large field of view of the camera allows recovering the vehicle orientation and position even in presence of significant odometric errors. Like most navigation systems, ROBOLIFT combines dead reckoning with a

Figure 4. ROBOLIFT, autonomous fork-lift by vision control.

discrete sequence of external measurements of the vehicle pose to correct cumulative errors.

In fact the odometric values are affected by errors (due to slipping wheels or small deformations and irregularities), and these errors tend to increase with the amount of the distance covered and the complexity of the maneuvers.

The 3-D self-localisation of the vehicle is obtained with a camera, suitably installed on the vehicle (as shown in Fig. 4) in order to frame the artificial H-shaped landmarks along the planned navigation route. Figure 5 refers an example of the landmark as it appears when recorded by the camera, which depends on the acquisition constraints (height and orientation of the sensor with respect to the navigation floor).

The visual navigation control performs model-based vision recognition, which is strongly driven by the knowledge of the shape and the symmetries of the landmark. Again, attentive vision mechanisms are used to find, quickly and reliably, all candidates of the geometrical features of the landmark. This is obtained by scanning only a subset of the image rows and looking for edge transitions and local symmetry constraints. Then, the detected shape contours recorded by the image sensor are projected onto the floor where geometric reasoning is performed to avoid any ambiguity with other structures, which may be present in the scene (see Fig. 6).

The performance of the vision system is strongly conditioned by the accuracy of the camera calibration process, which allows to compute the actual position and orientation of the camera sensor with respect to the odometry of the vehicle, to obtain a homogeneous data representation for navigation.

Figure 5. The image of a landmark as acquired by the navigation camera.

Figure 6. The result of the landmark detection algorithm. The 3D model of the landmark is projected onto the floor plane and appears superimposed to the original image.

CONCLUSIONS

The use of biological models in computer vision is not only aimed to achieve a better understanding of visual perception. Industrial applications make often use of hints coming from biological visual system studies, to get improved performance.

Attentive vision mechanisms are often exploited to reduce the amount of data to be processed and concentrate all available processing power to focus onto selected processing areas and specific detection objectives. Another fundamental requirement is system reliability and robustness to external disturbing events.

The chapter has briefly examined the main approaches currently investigated in the scientific community, namely the bottom-up saliency features scheme and the top-down model based approach. From an industrial viewpoint a full bottom-up approach is still unacceptable, unless a deep engineering control of all consequences and side effects could be fully supported.

The most promising and widely accepted solution turns out to be a mix-mode approach as a combination of low-level feature detection followed by local analysis with as much as possible a priori information. It is very important to make use of alerting mechanisms based on low-level features invariance properties. High-level strategies are always required to drive the understanding process, including context constraints.

The system should be able to achieve progressive learning capabilities in order to refine and improve the available models, including all typical variability (statistical and structural) which cannot be fully caught by the a priori modelling scheme.

REFERENCES

1. D.H. Hubel, *Eye, Brain and Vision*, Scientific American Library, by Scientific American Books Inc., New York (1988).

2. D. Marr, *A Computational Investigation into the Human Representation and Processing of Visual Information*, W.H. Freeman & Co., San Francisco (1982).

3. D.H. Ballard and C.M. Brown, *Computer Vision*, Prentice-Hall, Englewood Cliffs, New Jersey (1982).

4. http://www.klab.caltech.edu/~itti/attention/

5. *Proceedings of the Advanced Technology Conference*, USPS (1992).

6. C. Thorpe and M. Hebert, Mobile Robotics: Perspectives and Realities, *Proceedings of ICAR'95*, Saint Feliu, Spain, Sept. 1995.

7. E.D. Dickmanns, R. Behringer, D. Dickmanns, T. Hildebrandt, M. Maurer, F. Thomanek, and J. Schielen, The Seeing Passenger Car VaMoRs-P, *IEEE Symposium on Intelligent Vehicles '94*, Paris, pp.68-73, Oct. 1994.

8. G. Garibotto, ROBOLIFT: vision guided autonomous fork-lift, *Service Robot: An International Journal*, Vol 2 n.3, MCB University Press, pp.31-36, (1996).

PURPOSIVE VISUAL PERCEPTION AND CO-OPERATIVE BEHAVIOUR: SOME ISSUES FOR THE DESIGN OF PHYSICAL AGENTS

Giovanni Adorni, Stefano Cagnoni and Monica Mordonini

Dipartimento di Ingegneria dell'Informazione
University of Parma
Parco Area delle Scienze 181A
43100 Parma - Italy

INTRODUCTION

Looking back at the history of computer vision over recent years we can see how this area of research has evolved. Initially, in the early seventies, researchers were essentially involved in image processing. This mainly involved image transformations aimed at enhancing image quality and detecting image features. During the eighties, researchers turned their attention to the recovery of the three-dimensional scene information derived from two-dimensional images, based on geometric and photometric models of the imaging process. The nineties have been mainly characterised by the study of the integration of image analysis and synthesis to produce versatile visualisation techniques, namely *image media processing*, and by the integration of visual perception and purposive action to perform intelligent tasks, namely *purposive vision.*[1]

Among others, the MOB-LAB project has been a significant example of a purposive vision system, working on a "no-toy" world.[2] MOB-LAB is a MOBile LABoratory, developed during the seven-year European EUREKA-PROMETHEUS project, which develops and tests real-time intelligent systems, that monitor the trajectory of a vehicle by means of computer vision systems, with the possible inclusion of alerting the driver in dangerous situations. MOB-LAB demonstrated good performances running on extra-urban roads in the following tasks: obstacle and road detection, lane tracking control, overtaking control, estimation of the safety distance. Although MOB-LAB has proved to be an important project with very interesting results where real, complex, dynamic scenes are concerned, it suffers from what Bobrow called the "isolation assumption" of artificial intelligence.[3] We believe that a significant challenge for artificial intelligence at the end of this millennium is *to design intelligent systems that can co-operate purposively* with one another, possibly with humans, and with the real physical world (we call these open intelligent systems).

An *open intelligent system* is an intelligent entity whose intelligence does not reside solely in the "brain" of the system but emerges from the interaction with the operating

Human and Machine Perception 2: Emergence, Attention, and Creativity,
edited by Cantoni *et al.* Kluwer Academic / Plenum Publishers, New York, 1999.

environment and from the co-operation with other open intelligent systems through perception, action, and communication. Therefore, the integration of perception, action, and communication takes on a crucial role in open intelligent systems, and this requires the study and design of suitable architectures.

In this chapter we discuss the components of an architecture based on agent technology, suitable for the design of open intelligent systems, together with some preliminary results derived from its implementation on different physical platforms. The next section introduces the ideas on which the architecture is based. An application of the architecture to the design of a multi-agent autonomous robot navigating in an indoor environment is then discussed. It is illustrated an adaptation of the system previously described to a new "purpose": how to solve the auto-localisation problem of an autonomous robot. Finally it is presented a further application of the architecture to the design of a multi-agent robot soccer player.

AN ARCHITECTURE PROPOSAL

The artificial intelligence community has been conducting research in the field of intelligent agents (or rational agents) for more than a decade. To date no authoritative definition of an intelligent agent has been found, because different objectives and environments have led to the development of a variety of languages and architectures.[4] However, most researchers agree that a rational agent must have the following characteristics:[5-7] *autonomy*: it must be able to operate without direct, continuous supervision; *social ability*: it must be able to interact with other agents and, if possible, with humans; *reactivity*: it must be able to perceive the world in which it is acting and reacts to changes in a "just-in-time" manner; and *pro-activeness*: its behaviour must be not only reactive but it must be capable of taking initiatives. It is the combination of autonomous, social, reactive, and pro-active behaviours that distinguishes an agent from objects and actors.[8] An agent is an object that is able to reason and co-operates via structured messages. Agents encompass the behaviour of actors, in that they have their own thread of control.

As a consequence of the previous considerations, we can introduce a *Physical Agent* (PA) as a rational agent with visual perception, purposive action, and communication capabilities.

Intuitively, perception and purposive actions by a single PA can be modelled by the following state transition functions:

$\text{PERCEPTION}_i(k)$:

$$\text{WORLD\_STATE}(k) \times \text{INT\_STATE}_i(k) \rightarrow \text{PERCEPT}_i(k)$$

$\text{PURPOSIVE\_ACT}_i(k)$:

$$\text{INT\_STATE}_i(k) \rightarrow \text{INT\_STATE}_i(k+1) \times \text{WORLD\_STATE}(k+1)$$

where:

WORLD_STATE(k) denotes the state of the external world where PA_i, or possibly more than one PA, operate;

$\text{INT\_STATE}_i(k)$ denotes the internal state of PA_i. It includes different types of information: camera parameters (camera position, viewing angle, zooming factors, focus, etc.) and infrared and ultrasonic sensor parameters which effect perception; state of the physical body, state of the perceived world, common sense knowledge, etc.;

$PERCEPT_i(k)$ denotes entities (e.g., features, cues, markers, objects, etc.) perceived by PA_i at state "k".

A purposive action is a coordinated movement of a PA in a working environment performed in order to reach an objective.[9] Therefore, in modelling purposive actions we must consider the following issues; i) methods to generate coordinated movements of a PA (skills); ii) techniques for modelling a working environment (e.g., geometric and functional modelling) and for representing the common-sense knowledge about the physical world (e.g., qualitative physics); iii) methods to reach an objective (planning). A $PURPOSIVE\_ACT_i$ at the state "k" can modify its own INT_STATE_i as well as the WORLD_STATE bringing the system into a new state "k+1".

There are cases in which a purposive action performed by a PA_i implies the change of its INT_STATE_i without any side effect on the WORLD_STATE; that is the case, for example, of a surveillance system. We call such a PA an Observer Agent (OA). For an Observer Agent a purposive action is then defined as:

$$PURPOSIVE\_ACT^{OA}{}_i(k):$$
$$INT\_STATE_i(k) \rightarrow INT\_STATE_i(k+1)$$

Reasoning is the link between perception and purposive action obtained by means of some knowledge (e.g., knowledge about prototypical environments, common sense knowledge, etc.). In the case of a Physical Agent (and an Observer Agent) we can then define:

$$REASONING_i(k):$$
$$PERCEPT_i(k) \times INT\_STATE_i(k) \times KNOW \rightarrow INT\_STATE_i(k+1)$$

where:

KNOW denotes the knowledge of the system. It is worth noting that, at this stage of the research, we accept the hypothesis of "static" knowledge without any kind of learning capabilities.

As previously mentioned, communication capabilities are a key issue of a PA. COMMUNICATION can be defined by the following state transition functions:

$$SEND_i(k):$$
$$STATE_i(k) \rightarrow MESSAGE_j$$
$$RECEIVE_i(k):$$
$$MESSAGE_i \times INT\_STATE_i(k) \rightarrow INT\_STATE_i(k+1)$$

where:

$MESSAGE_i$, $MESSAGE_j$ denote messages sent out/received via the communication network.

It is worth noting that $SEND_i(k)$ does not depend on the receiver's state $STATE_j(k)$. In other words, it represents an asynchronous message transfer, and can support broadcasting. $RECEIVE_i(k)$ depends on the receiver's state $STATE_i(k)$. In other words, while a sender submits a message when it deems fit, a receiver can accept or reject the message depending on its own state.

It is also possible to define a Blind Agent (BA). This is a Physical Agent without perception; thus a pure purposive action executor which receives information about the external world by means of a communication network. For a Blind Agent, reasoning is defined as:

$$\text{REASONING}^{BA}_{i}(k):$$
$$\text{INT\_STATE}_i(k) \times \text{KNOW} \rightarrow \text{INT\_STATE}_i(k+1)$$

From the previous definitions we can denote a Soft Agent (SA) as a pure "reasoner" with communication capabilities. Figure 1 illustrates the general architecture of such an agent.

INDOOR VEHICLE NAVIGATION

The potential utility of machine vision for robot navigation in indoor environments has been under investigation since the early experiments with Shakey the robot at Stanford Robotics Institute.[10] Methodologies and vision algorithms for indoor navigation can exploit the peculiar properties of the domain. For practical situations, indoor set-ups are significantly more engineered than the outdoor counterparts. An office- or factory-like environment has a physical structure which is largely invariant over time and can be effectively described by appropriate topological and geometrical representations (maps). Building interiors and artefacts are typically regular in shape and of certain colour. These properties, combined with low electronic noise, greatly reduce the complexity of many vision algorithms. Of course, indoor mobile robots are not subject to weather conditions (for example, rain and fog) which can significantly impair vehicle visual perception. Unfortunately, even when applied to indoor navigation, vision suffers from some drawbacks; because of the relatively short distances involved, perspective deformation severely influences the perception of indoor scenes. Moreover, lighting conditions may be inconsistent because of the presence of windows and lights.

Many mobile robot systems rely only on ultrasonic and infrared sensors for their indoor navigation. Control techniques based on these sensors allow for robust obstacle

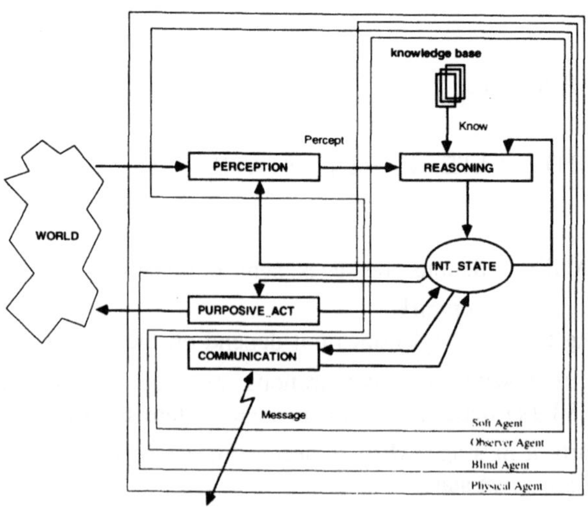

Figure 1. General architecture of an agent.

detection, wall following and even the building of topological maps. Within these systems, an on-board camera is possibly only present for the specific purpose of object-recognition and not for other navigation tasks.

Vision can constrain the robot path even though no explicit representation of the environment is developed. Mobile robot guidance is obtained through the use of visual features (for example, lines or tracks). For example, the vanishing points of straight lines (for example, ceiling or floor patterns) can be detected in 2-D images to provide robots with orientation information, but must be complemented by other sensors that are not sensitive to translation.[11]

If the robot is given a map of the environment a visual self-localisation process can be performed through the identification of known natural landmarks, artificial landmarks, vertical edges, which are also perceived through 360° omnidirectional image sensors. The main issues concern the choice of the optimal landmark shape or pattern needed to obtain robust localisation from single or multiple images and from the minimisation of position uncertainty.[12]

In indoor navigation the mobile robot must be able to identify objects which are part of its mission (task) representation: for instance, "go down the hallway and enter the last door on the right". Halls, doors, and other salient features in the environment must be perceived and located. The current robot position, on the other hand, is estimated integrating proprioceptive and heteroceptive sensors. A great deal of research work has thus been devoted to vision-based world modelling, which aims at constructing environment descriptions to be matched against their a-priori counterparts.

Simple visual features can provide an efficient and inexpensive constraint to mobile robot navigation in structured indoor environments. In an automated factory plant, automatically-guided vehicles follow predefined paths specified by the electro-magnetic field generated by hidden electric wires. Given the high costs involved in the maintenance and modification of automatically-guided vehicles routes, the use of painted tracks (highly contrasting the floor colour) as a guidance for camera-equipped mobile robots has been investigated (see, for example, Rajagopalan et al.[13]).

In the paragraphs that follow, we discuss some of the issues mentioned above from the point of view of the architecture illustrated. More precisely, we describe the components of a multi-agent system that allows a mobile robot to navigate in an indoor environment and, at the same time, solving the problem of robot auto-localisation. The robot is able to

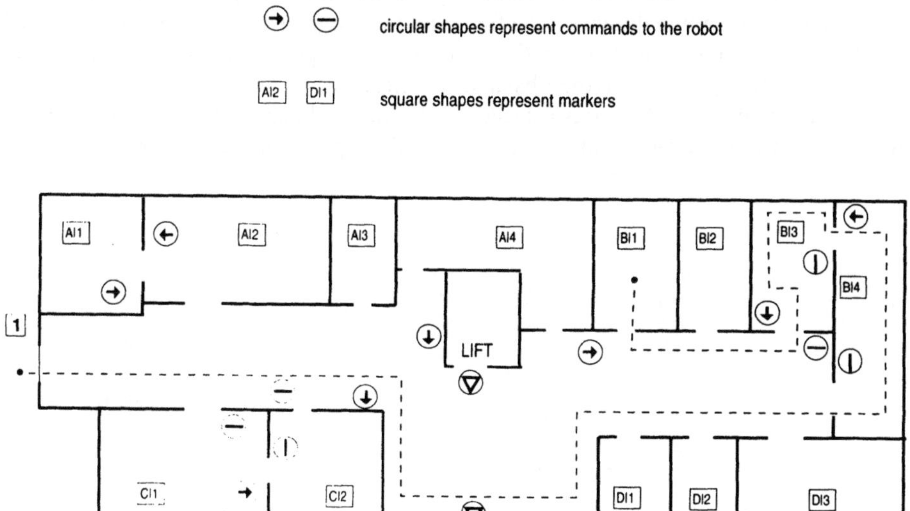

Figure 2. Layout of the experimental test bed.

perform the traditional goal-directed navigation task and to establish if there is a problem which cannot be solved (for instance, the robot has to enter a room and the door is closed), in which case it requires help from a human operator via a radio link. Moreover, when given a map of the location of markers in an indoor environment, the robot is also able to estimate its position from the information extracted from perceived images.

The robot navigates in a typical office environment with partitioned and communicating rooms. The robot has a topological knowledge of the indoor environment, some behavioural motor knowledge (for instance, "navigate along the median axis of the free space", "navigate close to the wall on your right", or "navigate following a track on the floor"), some knowledge of traffic signs and markers and of their typical position in the environment, and some knowledge of its own size.

The robot is asked to follow a series of instructions that tell it which room to enter. The instructions to the robot comprise simple statements such as "exit the room and turn left" or more complex statements as "go from point 1 to room B1 passing through room B4 applying a right wall following strategy" (see Fig. 2). Proceeding through the environment the robot, on the basis of its topological knowledge, looks for the markers (e.g., the name of the office) located on the doors, and for traffic signs which correspond to commands for the robot itself (for instance, a right arrow inside a circle indicates the command "turn approximately 90° to the right and continue with your navigation strategy"). In the case of navigation strategies based on free-space computation, traffic signs (if any) are located on the right side of a doorway if the sign contains a command indicating how to enter the office; they are located on the wall if the sign is a navigation command. In the case of navigating strategy driven by tracks on the floor, traffic signs are located on the floor, on the right side of the track (see Fig. 3). Traffic signs can change dynamically during a navigation task: for example, a traffic sign can be removed from the environment, it can be modified, its position can change or a new sign can be added to the environment.

During the navigation task the robot, equipped with a b/w CCD camera on a pan-tilt head and sonar sensors, uses sonars to compute the local free space and the CCD camera to follow the tracks and to recognise markers and signs or doors. Figure 4 shows the configuration of the robot used for the experiments described in this section.

The system comprises a Navigation Agent (which is an Observer Agent Type) allowing the mobile robot to move, and a Master Agent (which is a Soft Agent Type) that provides the Navigation Agent with a co-ordination plan to resolve any conflicts which it cannot solve by itself. Figure 5 shows an example where the Navigation Agent allows the mobile robot to follow a line drawn on the floor and to recognise traffic signs by means of a Visual Perception Agent (which is an Observer Agent Type). It also allows the mobile robot to detect, where possible, avoid obstacles by means of a Detector Agent (which is an Observer Agent Type). Traffic signs (placed on the floor), lines, and line intersections are identified and recognised by means of the Visual Perception Agent. Figure 6 shows an

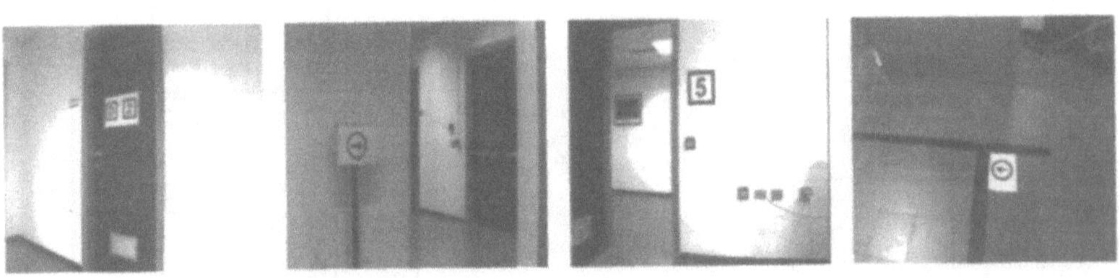

Figure 3. Examples of location of markers and traffic signs in the environment. From left to right: marker located on the doorway; traffic sign located on a pole; marker located on the wall; marker located on the floor on the right side of a track.

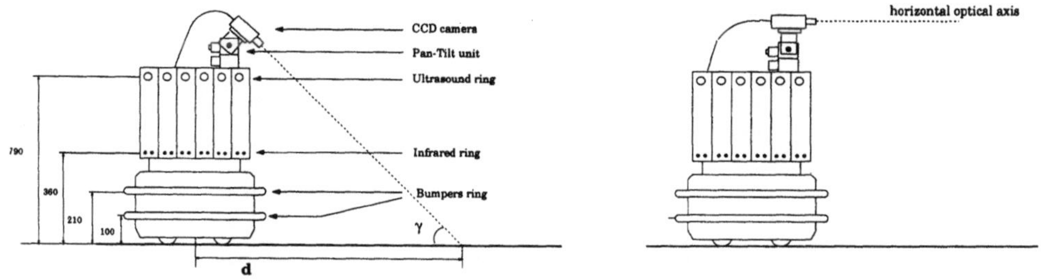

Figure 4. Camera configuration of the autonomous robot.

example of traffic signs used during the testing of the system. When the Visual Perception Agent recognises a traffic sign, then the Navigation Agent checks if the traffic sign implies a behavioural rule that modifies the planned actions. If this happens, the Navigation Agent modifies the current plan according to the new behavioural rule. Figures 6 illustrates some markers and traffic signs together with the corresponding behavioural rules. The Navigation Agent implements different "behaviours"; in the case of Figure 5, the applied behaviour is "follow a line".[14]

The Visual Perception Agent on board of a robot acquires images through a CCD camera, binarises them to reduce the amount of information to be processed, then extracts only the information necessary for line and intersection detection and sign recognition. All the targets to be identified are placed on the floor; then the pan-tilt head holding the CCD camera is moved by the Visual Perception Agent in a configuration, where the angle γ, ranging between 30° - 45°, depends on the height (h value) of the robot (see Fig. 4).

Robot navigation along a detected track is performed by means of the algorithm proposed in Thorpe.[15] The parameters required for this process are extracted from the line position in the acquired images.[16] This search is performed over a simplified binarised image in which the effects of perspective distortion, due to the camera configuration, are removed. In this approach the robot is also able to detect and follow the right track in the case of "broken" lines or lines corrupted by spots on the floor.

Line intersections are recognised by means of simple masks without removing the perspective effect; a more complex process is applied to recognise traffic signs. Such a process analyses binarised images looking for possible object candidates through the use of a combination of neural networks.

All networks are multi-layer perceptrons with one hidden layer, acting as classifiers. A classifier network divides the input space among the classes "associated" with the output units. The values associated with the output units can be considered as a "likelihood index" for the classes. The network associates the input pattern with the class associated depending on which unit provides the maximum output value. Each classifier has 169 input neurons that allow the network to perceive the world trough a view window of fixed size.

The entire recognition of a traffic sign takes place in two steps:

- identification and recognition of the external shape of a possible object candidate;
- recognition of a symbol inside the shape.

The size of the traffic sign as it appears in the image must be lower than an upper bound given by the size of the view window[17] but also greater than an (experimentally set) lower bound, to make recognition possible. The previous two-step process acts on images with different resolutions. In particular, the final step works at higher resolution to allow

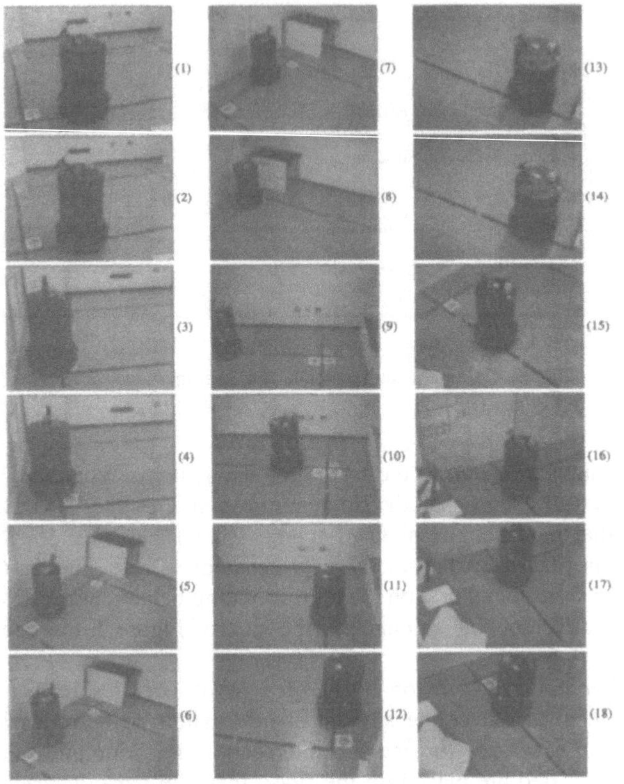

Figure 5. Example of indoor navigation by means of traffic signs following a track on the floor. During navigation, the robot recognises as commands only traffic signs placed on the right side of the track. In (1), (5), (9) and (12) the commands mean "turn right", while in (15) the command means "make a U turn".

for a correct recognition of the internal symbol. This procedure is applied to each part of a binarised image in which a possible candidate of a traffic sign has been detected.

To detect a possible candidate for the centre of a sign an attention-focussing algorithm is applied.[18] This algorithm is aimed at reducing the quantity of pixels to be processed by the neural networks, and to prevent incorrect patterns (patterns which are completely unknown), which may generate an unforeseeable network behaviour, from forming the network input. The analysed image is divided into four quadrants or leaves, each of which is recursively decomposed until it is elementary. An elementary leaf is: i) an image portion containing a number of black pixels lower or higher than two experimentally defined

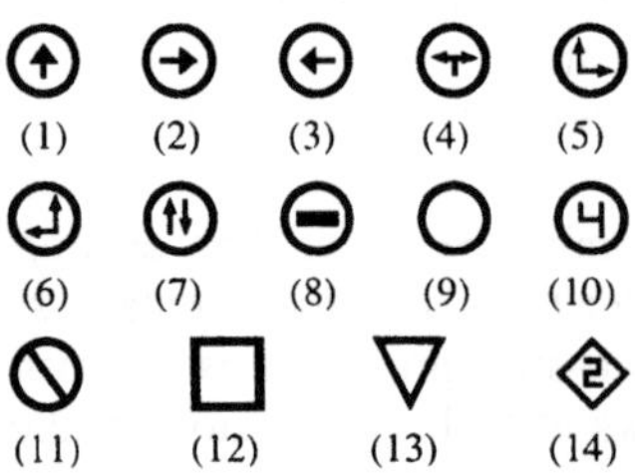

Sign	*mean*	*Sign*	*mean*
(1)	go ahead	(8)	no way
(2)	turn right	(9)	no thoroughfare
(3)	turn left	(10)	speed limit
(4)	turn right or left	(11)	end of speed limit
(5)	go ahead or turn right	(12)	take precedence
(6)	go ahead or turn left	(13)	give precedence
(7)	attention: alternate one way	(14)	segment marker

Figure 6. Set of traffic signs used for indoor navigation together with the corresponding behavioural rules.

thresholds (i.e., 7% of black pixels for the lower threshold and 30% of black pixels for the higher threshold, at lowest resolution); or ii) an image portion at lowest resolution (i.e., 16x16 in our experiment). The system looks for markers or signs (or portions of them) candidates on the leaves resulting from the previous steps, applying again the lower and higher thresholds. The result of this step is a portion of the original image of 16x16 pixels containing a possible marker or a sign candidate. The neural networks are applied to the image portion centred in the pixel selected in the previous step. The first classifier is able to discern among five classes: one for each shape depicted in Figure 6, plus one for the null class. The output of this step activates a second classifier that is able to recognise the internal symbol of the sign among the classes considered.

Network training has been performed on an appropriate set of external shapes and internal symbols, obtained by digitising and subsampling real images acquired by a robot CCD camera. To train the first classifier about 200 examples have been used for each shape class, including some negative examples. The training of each internal symbol classifier required about 150 samples for each symbol. Examples in the training set have been chosen so as to include most possible geometrical distortions of signs in the experimental environment.

Our tests resulted in a recognition success rate of 98% for the external shapes and of 99.2% for the internal symbols. The whole process is able to recognise correctly whole signs with a 99% success rate.

Obstacle detection is performed by the Detector Agent on the basis of data received from infrared and ultrasonic sensors. The infrared sensors that were used can, in theory, detect obstacles in the range 0.0 - 0.9 m, but their performance is heavily dependent, in the same lighting conditions, on colour (e.g., a black object is recognised only if its distance is less than 0.6 m). This limits the use of the infrared sensors to the range 0.2 - 0.7 m (it is worth noting that obstacle closer than 0.3 m may be located in the blind area between two different sensors).

Infrared sensors are always active. They detect obstacles near the robot along each direction, allowing the robot to avoid them or to stop. The detection of more distant obstacles is performed using the ultrasonic sensors.

The main problem related to the use of ultrasonic sensors is the presence of echoes. To overcome the echo effect, the "looking cycle" is repeated three times according to the following strategy: "if the results are similar, the detected obstacle distance is assumed to be their average, otherwise the smaller value is taken as the actual distance". To compute the direction of motion in the presence of obstacles, the robot performs the looking cycle twice: the first cycle is necessary to detect an obstacle (if any) found in a particular direction; the position of this obstacle after a second cycle allows the robot to establish if the obstacle is a moving object (possibly another robot or a human) and which direction it is moving. Experimental results demonstrated that the Detector Agent allows the detection of the obstacle position with a precision of 0.1 m.

Figure 7 shows an example of a different behaviour implemented by the Navigation Agent, that is, "goal oriented navigation task in the middle of the free space". In this example, the robot has to go from the present position in front of the marker (2) and then to go in front of the marker (5) avoiding obstacles. The configuration of the camera is in this case with the optical axis parallel to the ground; while obstacle detection is performed by the Detector Agent.

The duties of the Master Agent are very simple in the case when a single robot is present in the environment. The situation when a number of robots are present (see Adorni et al.[14] for a discussion of this point) is more complex. The Master Agent knows all the plans of the Navigation Agents of the different robots and the line segments where the mobile robots are expected to be. This is because the Navigation Agents send their plan to the Master Agent before starting their execution and communicate the new robot position

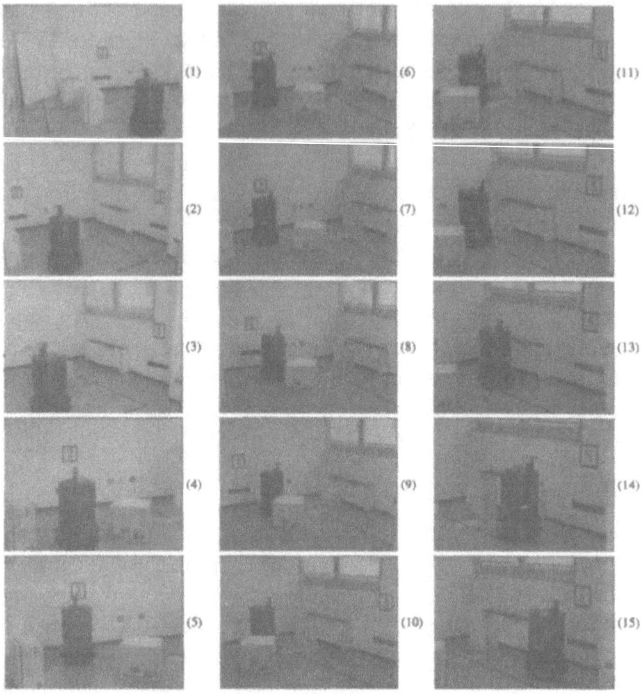

Figure 7. Example of indoor goal directed navigation task. The robot has to "go from the present position in front of the marker (2) and then go in front of the marker (5) avoiding obstacles".

when the previously planned trajectory changes. The coordination plan provided by the Master Agent contains motion and synchronisation actions. In particular, synchronisation actions correspond to the exchange of messages between the Master Agent and a Navigation Agent via a radio link.

ROBOT AUTO-LOCALIZATION

The architecture previously described also led to the development of a system able to solve the self-positioning problem of an autonomous robot.[19] Such a system is a mutation of that described above where we adapted the Visual Perception Agent and the Navigation Agent to the new different "purpose".

In this experiment, alphanumeric signs are placed in the robot environment as position landmarks (see Fig. 3) and perceived through an onboard CCD camera on a pan-tilt head able to scan a 270° horizontal view range; in addition, a topological representation of the environment, which also includes knowledge of markers location, is given to the robot.

Through the analysis of the perceived images the Visual Perception Agent must be able to estimate its position when it sees at least two signs. When the robot has to estimate its position, it stops and executes the following steps: i) movement of its pan-tilt head to search for a candidate to be a possible landmark; ii) recognition of an alphanumeric sign found within the previous candidate; iii) extraction of localisation parameter, if there is a maker; iv) application of the localisation procedure if the recognised markers are in a sufficient number, otherwise search for a new marker.

For a visual-based positioning method to be useful in practical situations, landmark recognition should be attainable over a sufficiently wide range of distances from the robot.

A straight application of the before mentioned visual perception capabilities makes it possible for the robot to identify markers of known dimensions in a range of 1.5-3 m. Since this range is not sufficient to localise a robot in a common like-office building, it has been necessary to modify the Visual Perception Agent adding a new screener (the focus of attention method) to apply to the image before the identification block (the same used in navigation task).[18] The goal of this different focus of attention is to give the robot the capability to recognise signs of variable size or the same sign at variable distances.

To solve this problem the robot has to exploit a common size-independent feature of signs: each sign is a closed shape with a connected internal shape. It is based on the recognition of different *connected regions* in the binary image, each connected region is labelled and its limits and centre coordinates are computed. Since a sign is composed of two connected regions having ideally the same centre and a well-defined size ratio, all the pixels that are, within reasonable error limitations, the centres of two regions at the same time must be examined, since they are sign "candidates". When these regions have the appropriate size ratio between external and internal connected regions there is a very high probability that the considered pixel under examination is the centre of a sign. The selected region containing a sign candidate has to be appropriately re-scaled before applying the neural networks, to allow the shape to be contained within the network "view window".

The parameters coming from a perceived image are two: the distance between the robot and the marker, and their angular displacement. Given the fixed marker size, if the camera is suitably calibrated, the distance can be easily computed (over a suitable range) by considering the relationship between the length in pixels of the fixed-size marker and its distance from the camera. In order to avoid possible weighting errors in the measurement, the image binarisation phase has to be designed carefully. The measurement accuracy depends also on image resolution, and robot-marker distance.

An increment of resolution decreases the error due to the spatial quantisation, which is directly proportional to robot-marker distance. Operating at a resolution of 256x256 pixels for distances ranging from 1 m to 7 m the maximum relative error is about 5% of the measurement distance.

Marker angular displacement α is the sum of two different contributions (see Fig. 8–left). The first is the current camera rotation angle α_{pan} as obtained from the pan-tilt unit controller. The second is the angle α_{offset} between the centre of marker and the centre of image, which can be reconstructed from the distance "h" between the two points in the image space and $d_{R,M}$ and it is known with a lower accuracy than the first. Only little angles of offset are accepted in order to decrease errors due to camera distortions and in these cases the angle measurement accuracy is less "sensitive" to the marker-robot distance value than to the distance error and it is limited to $\pm$ 0.03 rad.

Since the robot has to see at least two markers to localise itself, the robot is located on two circumferences: one centred in "M1" and radius equal to $d_{R,M1}$ and the second centred in "M2" and radius equal to $d_{R,M2}$ (see Fig. 8-right). By estimating the two relative distances $d_{R,M1}$ and $d_{R,M2}$, the robot can determine the intersection of the two previous circumferences. Disambiguation between the two possible positions (P1 and P2) is provided by the value of the angle α. This is the first method applied to estimate the robot positioning but it has a drawback: it may not provide a solution because of the errors made in the determination of the two distances. Each distance is known as $d_{R,M} \pm \delta\, d_{R,M}$, so the robot position has to belong to the intersection between two annuli centred in "M1" and "M2", with the lower radius equal respectively to $d_{R,M1} - \delta\, d_{R,M1}$ and $d_{R,M2} - \delta\, d_{R,M2}$ and the upper radius equal to $d_{R,M1} + \delta\, d_{R,M1}$ and $d_{R,M2} + \delta\, d_{R,M2}$.

Knowing the distance between the two seen markers $d_{M1,M2}$ in Figure 8-right and the angle between its direction and the y axis ($\phi 1$), through some simple geometrical constructions, it is possible to find the robot co-ordinates Rx and Ry in the world system, each of which is known with an accuracy proportional to the accuracy of the two distances. The value of these uncertainties determines a rectangular area in which we choose only the points belonging to the previous annuli intersections. To limit the uncertainty in Rx and Ry the nearest marker has to be chosen as reference marker. In the case of $d_{R,M1} << d_{R,M2}$ it is possible to compute $d_{R,M2}$ from $d_{R,M1}$ and $\gamma = |\beta - \alpha|$ to obtain a small area of uncertainty, which is a function of $d_{R,M1}$ and of the two angles: β and α.

With this method we easily obtain a smaller area of uncertainty for the robot position than with the simple application of the first method. The average error of each co-ordinate is about 10 cm, while the highest rectangular area is about 24x24 cm. The hallway of the ground floor of our building hosted several localisation experiments resulting in an average position error close to 10 cm. To improve the measurement accuracy the robot could use its onboard compass if the marker co-ordinate system is the same as for the compass. This means that the robot recognises the angle shown in Figure 8-right as τ. The reduction in measurement uncertainty is 3-4 cm for an error of about 10 cm in each robot co-ordinate value. In some situations the robot can see three markers and it could use this additional information to improve the accuracy in its localisation through a triangulation algorithm.[19]

The accuracy achieved with this method is higher only in those cases in which the markers are at a significant distance from the robot, but the uncertainty area seems to be less dependent on the distances because the angle measure error is less sensitive to the distance. In addition, we know not only three markers but also three distances and the robot has to be in the intersections of three annuli: this can limit the previous uncertainty area.

The Navigation Agent can implement any of the "behaviours" mentioned in the previous section; while the Detector Agent and the Master Agent remain essentially the same.

ROBOT SOCCER PLAYER

The RoboCup competition[20] is aimed at fostering an interdisciplinary approach to robotics and agent-based artificial intelligence, presenting a challenge in which teams of

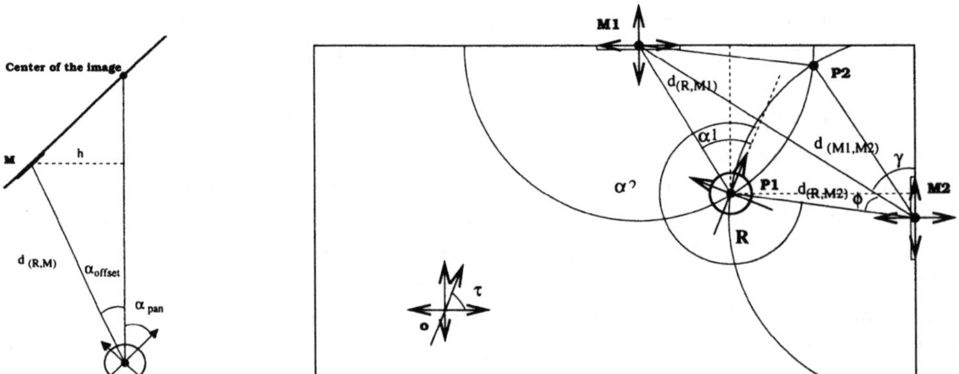

Figure 8. Geometrical construction of marker angular displacement (left), and robot positioning obtained from robot-marker distances and angular displacements.

"real" (small or middle size robot leagues) or "virtual" robots (simulator league) play soccer against one another.

In the simulator league, each team comprises eleven software clients, in communication with a central server. The server reconstructs the game, according to the commands received from the players. The server, in turn, sends back information to the players about their position, ball position and other environmental conditions that may affect the players' decisions and game strategy. In the small size league, competition is between teams of five small robots, which are controlled by an external program relying on a global vision system. The playing field size is as large as that used for table-tennis. In the middle size league, teams of up to four fully-autonomous robots (neither central global vision system nor external co-ordination systems are allowed) play on a field with a size of about nine meters lengthwise and five meters wide, with goals two meters wide. Robot dimensions cannot exceed 0.45 meters for square footprints or a diameter of 0.5 meters for circular footprints.

In every middle size team at least two kind of players must be present: a goalkeeper and up to three kickers that, in turn, can have their own specialisations. In the case of a goalkeeper, the robot visual field should be as wide as possible, since it has to be directed towards the centre of the playing field, while, at the same time, also keeping the part of the playing field on either side of the goal under control. For this reason, the goalie may also have a crucial role in team co-ordination, since it is the only player that can see almost all the field. Therefore, it can provide team-mates with global information about the location of the ball, opponents, and about the relative positions of team-mates.

The goalie's role and position inside the field strongly influence and constrain the design of its motion strategy. First of all, its movements must be aimed more at intercepting the ball rather than reaching or kicking it, since it must keep close to the goal. So, the goalie will move mainly sideways in front of its goal. Then, its movements must not extend beyond the goal boundaries, since it has to stand in front of it to defend it. Therefore, an interaction with the sensory sub-system must be provided in order for the goalie to be aware of its position relative to the goal.

The goalie must also be able to reposition itself independently, without relying on the data coming from its odometers. In fact, in cases of misalignment due to sudden contacts with other players, the encoders that transduce the wheel movements cannot record such a displacement properly. Moreover, it has to be able to perform short and rapid movements, to be able to follow even sharp deviations of the ball from its trajectory, due, for example, to a kick or to a rebound on the "body" of one of the players. The motion-control strategy must, therefore, enable the robot to react promptly to sudden changes, i.e., to exhibit a mostly reactive behaviour. The ball speed is relatively high and most computation time must be dedicated to processing the images coming from the camera system. For this reason, the specifications for the motion-control algorithm are that the input data, encoding ball position, speed and direction, must be concise and output commands must be as direct and immediately executable as possible. Finally the robot must be provided with communication capabilities, to be able to exchange messages with the other players on its team.

These constraints require that the physical layout of the goalie be considerably different from the other players' structure. Figure 9 shows the goalie prototype we have developed to tackle such issues,[21] where the two wide-angle cameras are on the top. The front wheel, mounted in the middle of the chassis, is partially hidden by the pneumatic kick device, whose air tank is visible just above it. The onboard additional hardware consists of a PC mainboard equipped with an AMD K6 200 MHz processor, a radio-link for communication, two frame grabbers for video acquisition.

As concerns the robot mechanical architecture, the two driving wheels on which the hardware platform we have used (a Pioneer robot[22] controlled by a Motorola 68HC11-based card that communicates with the mainboard through the serial port) is based, are

Figure 9. The goalie prototype.

located in the middle of the chassis, one on the front and the other on the rear. This makes translational movements more precise and accidental turns less likely. Balance is ensured by a pair of spheres, on which the robot leans, that are positioned along an axis at 90° to the wheel axis, and passing through its centre. Turning is possible because the two wheels can be operated independently.

The goalkeeper software architecture consists of a multi-agent system composed by three agents: a Visual Perception Agent (which is an Observer Agent Type), that provides the robot with vision; a Goal-Keeping Agent (which is a Blind Agent Type), that provides decision-making skills and controls the motion of the goalie; and a Cooperative Agent (which is a Blind Agent Type), which enables the robot to co-operate with the other robot players and co-ordinate their activities.

The Visual Perception Agent's duties are to detect and localise the objects that are in the playing environment. These can be distinguished by their different colours: the ball (red), the field (green with white lines), the sidewalls (white), the other players (black with a purple or light-blue mark on them, depending on the team on which they are playing), the two goals (one blue and one yellow). Such an agent's operations rely on a vision system based on two wide-angle cameras, that allow the visual field to be more than 180° wide.

The Visual Perception Agent implements three main behaviours: i) object detection and localisation; ii) absolute auto-localisation; and iii) self-repositioning. Object detection is performed through a colour-based segmentation algorithm. Such an algorithm returns the position of the ball and of possible other players occupying the goalie's field of view for each frame received from the cameras. Therefore, by comparing two subsequent frames and taking the robot motion into account, it is possible to deduce the instantaneous speed and direction of the ball, with respect to the goalie's position.

As it is not possible to acquire the images coming from the two frame grabbers at the same time (while the robot is moving), there is a switching strategy based on the detection of the ball, which is the only red object in the scene. The two fields of view of the two cameras overlap in a region which roughly corresponds to the centre of the zone that lies immediately in front of the goalie. When the ball is not detected by either of the two cameras, the camera from which the image is acquired is switched. When the ball is detected by one of the two cameras the image is acquired from such a camera until it enters the overlap region. At that point a switch to the other camera is performed.

Auto-localisation and repositioning are based on an algorithm which detects the lines of the goalkeeper area and infers the goalie's position through a comparison between the two images acquired by the two cameras. The goalie's area front line is first detected in the two frames. If the goalie is aligned with such a line, the two resulting images are symmetric. If the goalie is rotated, such a symmetry is lost. Figure 10 shows the field as seen by the goalie during a repositioning maneuvre, along with a diagram in which its motion is shown. The goalie starts from a position in which it is rotated by 45° on the left and is in the middle of the goalie's area. It then reaches the front line of the area and aligns itself with it.

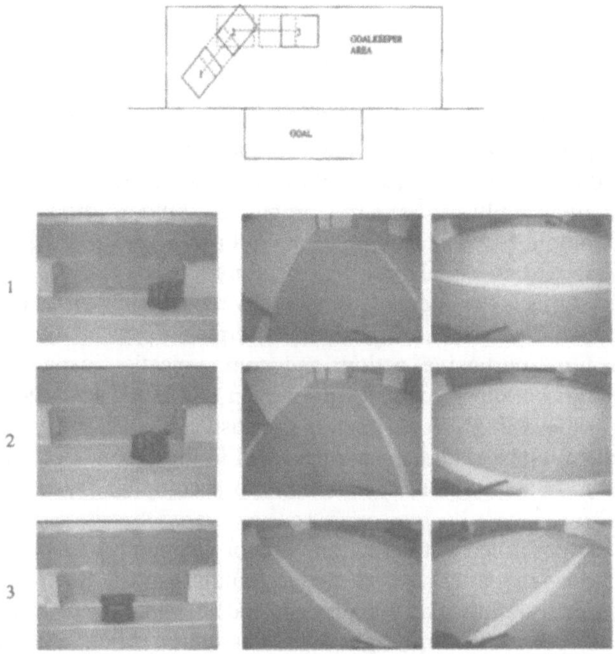

Figure 10. Example of the goalie's repositioning maneuvre. Top: scheme of the maneuvre; Left: three frames from the maneuvre sequence as seen from outside; Right: corresponding frames pairs as seen by means of the goalkeeper's left and right cameras.

The Co-operative Agent makes it possible for the robot to 'talk', using a messaging system, with the other team components and with a remote system from which, however, it is only allowed to receive a start signal at the beginning of each half. The Goal-Keeping Agent plans the robot behaviours according to the activity of the other two agents. In particular it drives the engines and actuators of the robot, and activates the robot behaviours in the different phases of the game.

The scheduling of the goalie's behaviours is essentially driven by the ball position. If the ball is close, a ball-blocking behaviour is activated; if the ball is far from the goalie or is moving away from him, self-localisation and self-repositioning (if necessary) are activated. The ball-blocking behaviour has been designed using genetic programming.[21]

A robot soccer player can have a hardware architecture which is quite similar to the goalie's. However it must differ from the goalie in its mechanics. The player's motion is quite different from the goalie's, as it has to be able to move in any direction of the playing field, to turn around in small spaces and to interact with the ball in a very different way. So a common solution would be a wheels layout using a couple of opposite driving wheels on one end of the player's chassis and a central free wheel, capable of rotating around an axis which is perpendicular to the ground, in the middle of the other end.

The software agent-based architecture can be the same as the goalie's. However, the main differences would be on the robot behaviours and communications. The Visual Perception Agent of the robot player must provide similar capabilities to the goalie's, even if their "purposes" must be different goals and based on different external reference structures.

CONCLUDING REMARKS

Intelligent Agents (also called Rational Agents) are an extremely lively area of research and application in artificial intelligence. Agent-based systems arise out of the following needs: adaptable, fault-tolerant distributed systems that solve complex problems; open systems where components come and go and new components can be added; migration and load balancing across platforms, through a network; new metaphors, such as negotiation, for solving distributed, multi-disciplinary problems. Among other characteristics, an intelligent agent is: autonomous, social, reactive, and pro-active. An Agent is an "object" that is able to reason and to collaborate via structured messages. Agents encompass the behaviour of "active objects" or "actors", in that they have their own control lines.

In this chapter we discussed some issues concerning *Physical Agents*, rational agents with visual perception, purposive action, and communication capabilities. A Physical Agent is an interesting "building block" for the design of *Open Intelligent Systems*, in other words, intelligent systems that can co-operate purposively with one another and with the very physical world. We tested the ideas introduced in the second section of this chapter, designing three different autonomous systems able to accomplish three different purposive actions.

The first system, described in the section "Indoor Vehicle Navigation", is a distributed multi-agent autonomous robot designed to navigate autonomously in an indoor environment on the basis of the interpretation of markers and traffic signs, by means of vision. Such a multi-agent system comprises three agents on board the robot, and an external agent connected via radio link to solve conflicts which the robot cannot solve by itself.

The second system, described in the section "Robot Auto-Localisation", is a distributed multi-agent autonomous robot which can solve the problem of robot auto-localisation by means of vision. Such a multi-agent system is derived from the previous one, where the vision system (Visual Perception Agent) has been modified to take into account the different "purpose".

The last system, described in the section "Robot Soccer Player", is a distributed multi-agent autonomous robot which can play soccer in the role of goalkeeper in the RoboCup competition. Such a multi-agent system comprises two agents on board the robot, and an external agent connected via radio link which enables the robot to co-operate with the other robot players and to co-ordinate their activities. The goalkeeper is a component of the ART - Azzurra Robot Team, the Italian team which took part in the world competition RoboCup'98 in Paris, and will be involved in future RoboCup initiatives.

From a scientific point of view, we proposed a functional dependency model to explore the mechanisms of the integration of perception, purposive action, and communication. From the technological point of view, we developed various algorithms for real-time purposive vision. Experimental results have shown that they can be used in the real world.

Further developments of the research are aimed at studying and designing multi-agent, multi-robots systems, to realise more complex co-operative and competitive behaviours, and at introducing temporal features to make the model really meaningful.

ACKNOWLEDGEMENTS

This work has been partially supported by the National Research Council (CNR) through the "Progetto Finalizzato Trasporti 2", and through the "Progetto Speciale SIAD".

We wish to thank all members of the "Azzurra Robot Team 1999" for their precious and continuous co-operation in the development of the goalkeeper.

REFERENCES

1. Y. Aloimonos ed., Special Issue on Purposive, Qualitative, Active Vision, *CVGIP: Image Understanding*, Vol.56, No.1, pp.1-129, (1992).
2. G. Adorni, MOB-LAB: a MOBile LABoratory for the design and test of real-time intelligent system, PROMETHEUS Report BRI01094, PROMETHEUS Office, Daimler-Benz AG, Stuttgart, October 1994.
3. D.G. Bobrow, Dimensions of interaction: AAAI-90 presidential address, *AI Magazine*, Vol.12, No.3, pp.64-80, (1991).
4. M.J. Wooldrige and N.R. Jennings, Intelligent Agents, *ECAI-94 Workshop on Agent Theories, Architectures and Languages*, Lecture Notes in Artificial Inteligence, Springer-Verlag, Amsterdam, (1994).
5. Y. Shoham, Agent-Oriented Programming, *Artificial Inteligence*, Vol.60, No.1, pp.51-92, (1993).
6. G. Adorni and A. Poggi, Object-Oriented Programming Languages: a Natural Framework for Distributed Artificial Intelligence, in A. Kent and J.G. Williams eds., *Encyclopedia of Computer Science and Technology*, Vol.33, pp.229-244, Marcel Dekker Inc., New York, (1995).
7. M.J. Wooldrige and N.R. Jennings, Intelligent Agents: Theories and practice, *Knowledge Engineering Review*, Vol.10, No.12, (1995).
8. G. Agha, *A model of concurrent computation in distributed systems*, MIT Press, Cambridge: MA, (1986).
9. G. Adorni, S. Gaglio, L. Massone, P. Morasso, and R. Zaccaria, Cognitive modeling of Purposive Actions, in P. Morasso and V. Tagliasco eds., *Human Movement Understanding*, pp.215-242, Elsevier Science Publ. B.V., Amsterdam, (1986).
10. N.J. Nillson, Shakey the robot, SRI Technical Report 323, Stanford Robotic Institute, Stanford, CA, April 1984.
11. R. Schuster, N. Asari, and A. Bani-Hashemi, Steering a Robot with Vanishing Points, *IEEE Transaction on Robotics and Automation*, Vol.9, No.4, August 1993.
12. J. Borenstein, H.R. Everett and L. Feng, *Navigating Mobile Robots: Sensors and Techniques*, A.K.Peters. Ltd., Wellesley, MA, (1996).
13. R. Rajagopalan, R.M.H.Cheng and S. Lequoc, A Guidance Control Scheme for Accurate Track Following of AGVs, *Procs. IEEE Int. Conf. on Robotics and Automation (ICRA'92),* Nice, France, (1992).
14. G. Adorni, M. Mordonini, and A. Poggi, Autonomous Agents Coordination Through Traffic Signs and Rules, *Procs, of the IEEE Intelligent Transportation Systems*, pp.125-131, Boston, November 1997.
15. C. Thorpe, *Vision and Navigation the Carnegie Mellon Navlab*, Kluwer Academic Publ., Dordrecht, (1990).
16 D.H. Ballard and C.M. Brown, *Computer Vision*, Prentice Hall, Englewood Cliff, (1982).
17. G. Adorni, M. Gori, and M. Mordonini, Just-in-time Landmarks Recognition, *Real Time Imaging*, Vol.4, pp.3-10, (1998).
18. G. Adorni, G. Destri, and M. Mordonini, Indoor Vehicle Navigation by Means of Signs, *Procs. of the 1996 IEEE Intelligent Vehicles Symposium*, pp.76-81, Tokyo, September 1996.
19. G. Adorni, G. Destri, M. Mordonini, and F. Zanichelli, Robot Self-Localization by Means of Vision, *Procs. of the First Euromicro Workshop on Advanced Mobile Robots (EUROBOT'96),* pp.160-165, Kaiserslautern, Germany, October 1996.
20. H. Kitano, M. Asada, Y. Kuniyoshi, I. Noda, and E. Osawa, ROBOCUP: the robot world cup initiative, *Procs. of IJCAI'95 Workshop on Entertainment and AI/ALife*, Montreal, August 1995.
21. G. Adorni, S. Cagnoni, and M. Mordonini, Genetic Programming of a Goalkeeper Control Strategy for the RoboCup Middle Size Competition, In P. Nordin and R. Poli (eds.), *Genetic Programming: Procs. of EuroGP'99*, LNCS Series No. 1598, Springer Verlag, Berlin, (1999).
22. ActivMedia, Pioneer Operation Manual, 2nd edition, ActivMedia, (1998).

PANEL SUMMARY: **PLANNING AND PROBLEM SOLVING**

Panellists: Salvatore Gaglio (*chairperson*),[1] Maurizio Cardaci,[2]
Ingrid Isenhardt,[3] Dietrich Brandt,[3] and Giuseppe Strina[4]

[1]Centro di Studio sulle Reti di Elaboratori (C.E.R.E.)
Università di Palermo
I-90100 Palermo, Italy
[2]Dipartimento di Psicologia
Università di Palermo
I-90100 Palermo, Italy
[3]Department of Computer Science in Mechanical Engineering (HDZ/IMA)
University of Technology (RWTH)
Dennewartstr. 27, D-52068 Aachen, Germany
e-mail: brandt@hdz-ima.rwth-aachen.de
[4]Institute for Entrepreneurial Cybernetics (IfU e.V.)
Mülheim/Ruhr, Germany

PROBLEM SOLVING: THE STANDPOINT OF A PSYCHOLOGIST

(*M. Cardaci)*

I will sketch some ideas that have guided the psychological approaches to problem solving. Traditionally, psychologists have considered the ability to solve problems "one of the most important manifestations of human thinking".[1] Moreover, "problem" has been defined as a search that starts when we have a goal, but our habitual means are not sufficient to achieve it.

However, we should admit that problem solving isn't only a "mentalistic" property of human mind. Indeed, problem solving involves a wide range of activities shared by animals (for instance, apes), by artificial systems (for instance, computer programs for playing chess), and even by plants.[2]

Since problem solving involves flexible sequences of action leading an adaptive end, it should be distinguished from other types of adaptive behaviours already incorporated in the biological equipment of organisms. Perhaps, instinct is the most fundamental type of adaptive behaviour planned by a long biological evolution. Instinct provides animals for preformed means to copy their natural environments and basically consists in repertories of automatically elicited responses to specific circumstances.

By contrast, problem solving is similar to learning, i.e. to the acquisition of *new* abilities. Early psychologists hypothesised that problem solving, as gradual elimination of random mistakes, arises from trial and error learning. Even if this explanation could appear enough convincing, it does not consider that, in most situations, a "blind" search of the solution guided only by chance would lead the organism to insuperable difficulties, to time-consuming and to negative consequences of making errors.

Adopting a functionalist point of view, we should stress that, by definition, a "problem" doesn't exist per se, but it has to be discovered by the organism.[3] In other words, problem solving starts only when an interaction between organism and environment is experienced by the organism as a source of perturbation which demands a choice between several possible actions or the construction of new strategies. In extreme conditions, if the organism doesn't recognise the perturbation, it risks to be overcome. Therefore, we may assume that organisms have some kind of inner equilibration devices, probably based on action-perception and attention-memory loops. These mechanisms could not only detect the occurrence of a perturbation, but also activate a motivational drive to search a solution. Despite the motivational component of problem solving has been often neglected by several theories, motivation plays a fundamental role in planning and executing goal-oriented actions. Indeed, problem solving exists only if a motivational state drives the organism to anticipate and to maintain a goal, to plan goal-oriented activities, to predict and to monitor their effects, to correct its strategies on the basis of feedback.

In the last decades, the study of different varieties of cognitive strategies has led psychologists to describe problem solving as search in a metaphorical space or as Gestalt restructuring (or analogical mapping).

As regards the search metaphor,[4] problem solving can be viewed as a heuristic search (Means-Ends Analysis) or as a set of condition-action rules,[5] which allow an intelligent agent to run operations step-by-step. After having detected the distance between the current state and the goal state, the agent breaks an overall problem into specific subgoals which can be achieved separately. For any particular subgoal, the agent sequentially tries alternative operators in the search for a solution path. As in most conditions an extensive search in the solution space is computationally impossible to pursue, the core of an intelligent search is the use of heuristic methods which select a small number of salient choices that seem most likely to provide reasonably good, but not necessarily optimal solutions.[6]

However, we know that most practical problems not only require a massive computation, but also are often *ill-defined*[7]– they cannot be perfectly decomposed into independent parts that can be sequentially processed. Gestalt restructuring and analogical mapping provide us an approach to this kind of situations. According the Gestalt viewpoint, problem solving basically consists in a perceptual reorganisation of the cognitive field in a series of hierarchically related solutions which tend to become increasingly more specific. The final solution tends to occur as a sudden, unpredictable reorganisation of the perceptual field (insight) and can be transferred to similar situations.[8] The difficulty of problems is mostly related to the fixity of the perceptual field, that hinders the emerging of new insightful solutions. The Gestalt approach relies on parallel information processing and on parallel constraint satisfaction. The solution of each subgoal depends on the solution of other subgoals and a network of knowledge has to be constructed to reach a highly interactive solution. Gestalt restructuring is closely related to analogical mapping, because, as a number of researchers actually point out, analogy is a mechanism which could support insightful solutions and their transfer to similar problems. In analogical, case-based reasoning, the agent uses features of the target problem as retrieval cues to access a source analogue encoded into memory in a different context.

Then the problem solver tries to find a structural consistent mapping between the elements of the source and target. Mapping a set of correspondences between the elements of the two analogues may be constrained by perceptual similarities, by a conceptual isomorphism as well by searching similar goals in different situations.[9]

This brief outline should convince us that problem solving is not reducible to all-or-nothing strategies or to simple true-false operations. Recognising a problem and approaching it from different angles, generating heuristic or Gestalt strategies, imaging new possible relationships, grasping the early emergence of useful solutions, finding novel and original issues, are facets of creative activities which cannot be easily explained by a single comprehensive theory. Nevertheless, according Johnson-Laird and Byrne, problem solving, as other modes of thought, essentially demands for the *manipulation of models*. A model has a structure that is remote from verbal or propositional representations, but "close to the structure of the world as humans conceive it".[10]

At the end of my discussion, let me mention a great Gestalt-oriented psychologist, Karl Duncker who, on the basis of his investigations, described problem solving as an open process... *"what is really done in any solution of problems consists in formulating the problem more productively ...The final form of a solution is typically attained by way of mediating phases of the process, of which each one, in retrospect, posseses the character of solution, and, in prospect, that of a problem."*[11]

PLANNING AND PROBLEM SOLVING IN ARTIFICIAL INTELLIGENCE
(*S. Gaglio*)

In artificial intelligence a central concept is that of *intelligent agent.* An intelligent agent lives and performs given tasks in an environment, which can be either real or artificial. From the point of view of engineering, a significant relevant class of intelligent agents is that of rational agents.

Russell and Norvig[12] define a rational agent in the following way: "For each possible percept sequence, an ideal rational agent should do whatever action is expected to maximise its performance measure, on the basis of the evidence provided by the percept sequence and whatever built-in knowledge the agent has". This means that a rational agent must possess all the capabilities required to solve problems in an effective way.

Problems can be formulated in a formal way according to the state space paradigm: starting from an initial state and repeatedly applying a suitable successor function, which encodes all the possible simple actions, all the reachable states can be generated. Problem solving consists of building a search tree (see Fig. 1), which represents some possible

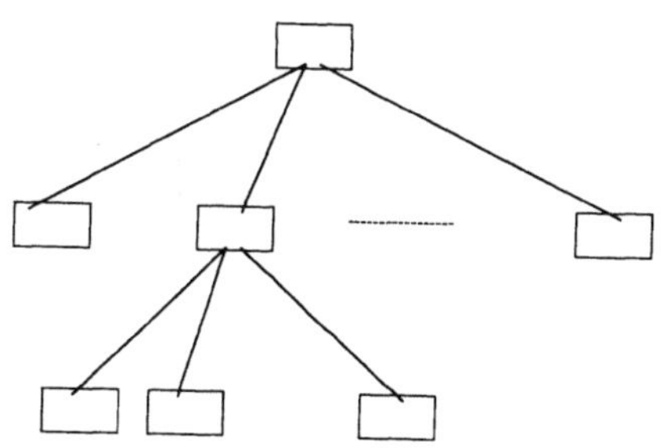

Figure 1. A search tree represents allowed transitions to new states according to possible actions, starting from an initial state.

sequences of actions, and finding in it a path that leads from the initial state to a state in which the required goal is achieved. Searching in a direction that goes from an initial state toward a goal state is called *forward search*. Alternatively, a *backward search* can be accomplished starting from a goal state and ending on the initial state, finding at each step all the possible previous states of a given state. This kind of search is also called goal-based. In any case a solution is a path on the search tree that corresponds to a sequence of actions.

The construction of the search tree can proceed according to classical general strategies like breadth-first, depth-first, or their variations or combinations, which are known to suffer of exponential growth at least in time. The use of a heuristic function speeds up the search and sometimes can achieve a sub-exponential growth of the tree. This strategy is known as best-first. The most popular algorithm belonging to this class is A* and it selects the next node to expand in the tree according to the following evaluation function:

$$f(\mathrm{n}) = g(\mathrm{n}) + h(\mathrm{n}) \tag{1}$$

which is the estimated cost of the cheapest solution though n, where $g(\mathrm{n})$ is known and is the cost of the path from the root to the node n, and $h(\mathrm{n})$ is the *heuristic function* which is an estimate of the cost of the cheapest path from the node n to a node representing a state in which the goal results satisfied.

In the artificial intelligence community, planning is a special approach to problem solving in which automated reasoning procedures over explicit logical representations for states, goals, and actions are used.

Present representation languages for planning derive from the STRIPS language introduced by Fikes and Nilsson[13] in 1971 for the STRIPS planning program, designed to control a robot named Shakey which had to navigate and perform simple actions in rooms and corridors of SRI International (see Fig. 2). The STRIPS language is based on First Order Logic. It represents *states* with formulas like

At(Shakey, x),
$\mathrm{In}(x, r) \wedge \mathrm{In}(y, r)$,
On(Shakey, Floor),

to mean respectively that Shakey is at the location x, the locations x and y are in the same room, and that Shakey is on the floor.

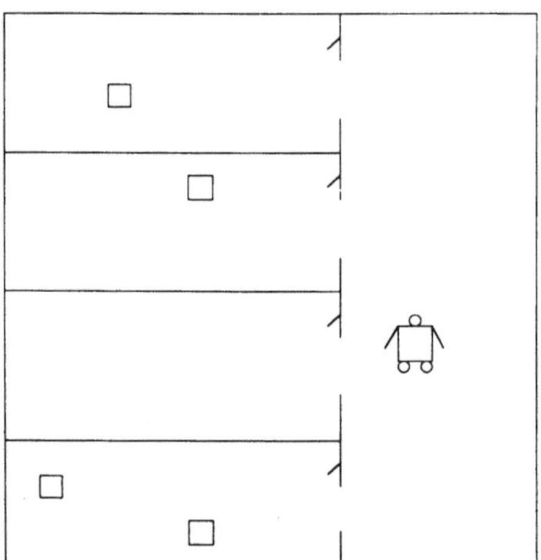

Figure 2. The Shakey's world made of rooms, corridors, boxes, and light switches.

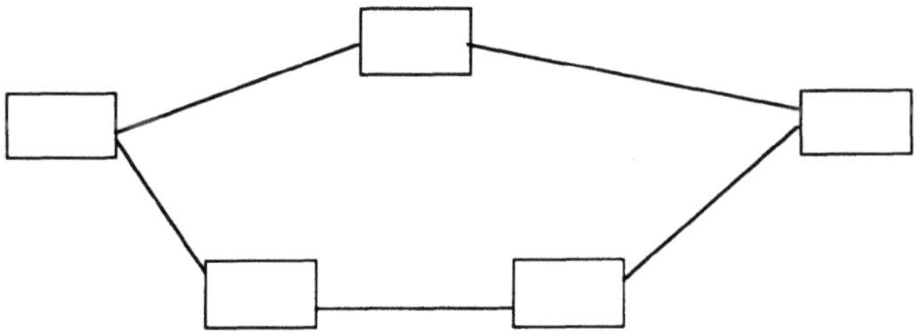

Figure 3. A plan is a partially ordered structure in which the time relationships among actions are made explicit.

Actions are represented in terms of their preconditions – necessary conditions to be applicable – and of their effects. For instance, the following action

Go(y)
Preconditions: At(Shakey, x) $\wedge$ (In(x, r) $\wedge$ In(y, r))
Effects: At(Shakey, y) $\wedge$ $\neg$At(Shakey, x)

encodes the fact that Shakey can go to the location y if it is already in a location x in the same room, while, after going, the effect is that in the new state it is at the location y and not any more at the location x.

Other actions related to the Shakey's world are Push(b, x, y), Climb(b), Down(b), TurnOn(ls), TurnOff(ls), etc., which mean respectively the pushing of the box b from x to y, the climbing on the box b, the turning on and off of a light switch, etc..

A plan can be built through a search on state space using problem solving algorithms. However, this turns out to be not practical because of the high branching factor of the search tree and of the huge size of the state space.

To overcome these difficulties, a possible approach consists of searching through a *space of plans*. Accordingly, the search starts with a *partial plan* and proceeds by applying *refinement operators* or *modification operators* until it achieves a *complete and consistent plan*. Plans are conveniently represented in terms of partially ordered structures, in which time relationships among actions are explicitly represented (see Fig. 3).

Unfortunately the real world is complex and not completely predictable, therefore different methods have been devised to build practical planners.

One approach is that of *hierarchical planning*, in which the planning activity proceeds at different levels of detail. Each action, therefore, can be in turn possibly expanded in a more detailed plan, as in Figure 4.

Another approach is that of conditional planning, in which suitable introduced sensing actions allow decisional statements like the following

If[Clear(Support), [PutOn(x, Support)],[Free(Support), PutOn(x, Support)]

Other approaches require *action* or *execution monitoring*, which can start replanning activities whenever unexpected events arise, or a continuous intermixing of planning and acting.

A complete different approach is that of *reactive planning*, in which the agent has not an explicit representation of the world. Instead, it possesses a set of preprogrammed behaviours organised according to defined preference relationships, as for instance in the *subsumption architecture* proposed by Brooks.[14] In the special case of robotics, Arkin[15]

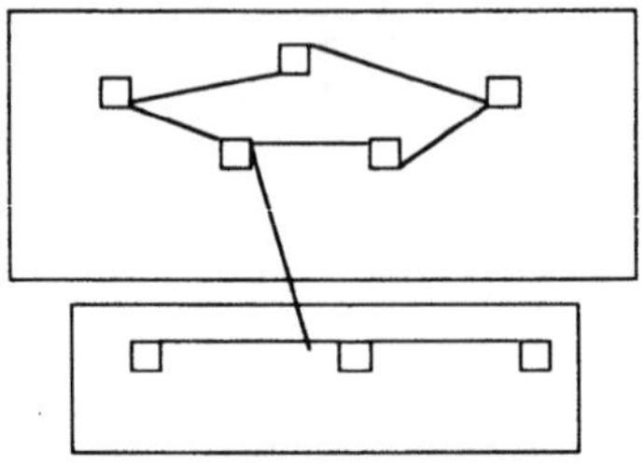

Figure 4. Expanding actions in hierarchical planning.

defines reactive control as a "technique for tightly coupling perception and action, typically in the context of motor behaviours, to produce timely robotic response in dynamic and unstructured worlds".

Finally, it is possible to cope with the uncertainty which characterises the sensor–action based interaction with the world by using statistical based decision procedures, like dynamic programming, Markov decision processes, dynamic belief networks, and dynamic decision networks.

CHANGING COMPLEX SOCIOTECHNICAL SYSTEMS

(*I. Isenhardt, D. Brandt, G. Strina*)

The Concepts of Fractals and Self-Similarity

Looking at processes around us today, the question arises how to understand what is going on. It is widely discussed how humans gain knowledge and understanding of processes, and the methods, tools and strategies of such human activities are developing fast within psychology, philosophy, and engineering. The complexity of the world around us, however, is growing much faster than all these tools are developing. Very few strategies try to cope with this ever-increasing complexity caused by technology, networking, globalisation etc..

The challenge for humans becomes even more obvious if *change processes* are taken into consideration as they are taking place within the complex systems around us, and if we try to influence such change processes.

In this chapter, a strategy is suggested to understand and influence such complex processes. The approach takes up recent concepts of chaos theory because of its affinity to the world today. As an example, highly complex, networked enterprises and their change processes are considered.

It is suggested to describe enterprises following the concept of sociotechnical systems. This concept includes the modelling of the enterprise as a self-similar system. Such systems are able to continuously change and develop in order to meet all requirements of their tasks. For this aim, the self-similarity of *structures* and processes need to be considered in parallel: the self-similarity of the *structure* of the enterprise with its departments and working groups as well as the self-similarity of the change *process*.

The human process of understanding such complex structures can be characterised by the following hypotheses:

- there is no real *border line* between observers and actors within such structures - understanding happens through acting within the system;
- there is no *objective* knowledge, all knowledge is arrived at through communication and negotiation.

On this background, the chapter describes the strategy of understanding system structures and change processes as follows.

The Sociotechnical Concept to Understand Systems

Let us consider complex networked enterprises today. Such enterprises are *sociotechnical systems.* As a method to analyse, redesign and monitor such sociotechnical systems, the OSTO approach has specifically been suggested. [16-18] OSTO stands for "open, socio-technical-economic system", i.e. the open system comprises social and technical as well as economic components. It builds on the socio-technical system theory which was considerably influenced by members of the London Tavistock Institute of Human Relations.

OSTO focuses on analysing work processes which are strongly inter-related. They are largely the result of a combination of high level of technology and human communication. OSTO understands systems as "living systems" (open cybernetic systems). Such systems include humans with their processes of work and life. Feedback processes stabilise and renew this open living system (see Fig. 5).

The system structure is characterised by different design elements which are dependent on each other in many complex ways. The main elements are the technical, social and organisational sub-systems: technology, people and organisation.

Self-Similarity and Change

Enterprises are highly complex systems in terms of specific dynamic interrelations between people, the organisation they work in and the technology they work with. As mentioned before, such complex social systems must continuously develop and redesign themselves to meet all the requirements of their tasks.

Such change processes are in themselves highly complex. Concepts of chaos theory may cautiously be applied to analyse and structure such change processes. The main concept to be used here is the concept of fractals or self-similarity. Fractals - or "holons" - are entities which are self-similar on an unlimited scale. If the scale, however, is limited these entities may merely be called "self-similar". (In the famous Mandelbrot structure, self-similarity occurs through all mathematical scales for every part of the borderline; any zooming-in on its segments leads again to similar structures or "fractals").

In change processes, this concept may be applied in two different ways:

- the *structure* of the organisation is to correspond to the fractal principle (self-similarity of *entities*);
- the *process* of innovation towards new structures is also to correspond to the fractal principle (self-similarity of *processes*).

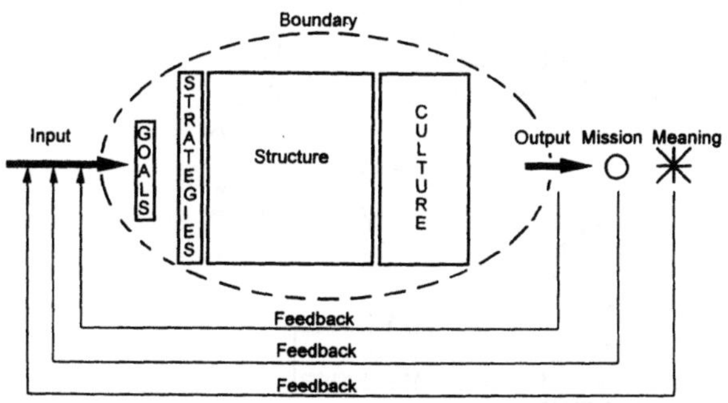

Figure 5. The OSTO approach: the sociotechnical system.

These two aspects are briefly described here:

a) *self-similarity of entities*: introducing group work in a company means to design decentralised sub-entities (e.g., working groups) where every sub-entity represents in a specific way the system as a whole. This concept corresponds to the present view of complex organisations which are to mirror within themselves the complexity and turbulences of their environment. Such organisations are only able to cope with high outside complexity if they are within themselves structured to cope with their own internal complexity. The competence of an enterprise to cope with complexity is greatly enhanced by using the "fractal" concept for its internal structures. It includes the competence to organise transitions between sub-systems (see Fig. 6);

b) *self-similarity of processes*: structuring the innovation process as a self-similar process means to define certain process steps which show up on all levels of complexity of this process. These basic steps of an innovation process may be defined e.g., as:
- planning of aims,
- developing action concept,
- designing and performing action,
- evaluation.

This process structure repeats itself in every process step, e.g., the planning of aims comprises: initial planning discussions, development of aims as a conceptual activity, defining and implementing the aims in a group consensus process (which can be understood as realising an action plan), and final evaluation of this step.

The main advantages of this self-similar *process* structure are as follows:

• the project phases are no longer separated from each other in a sequential order. The process appears more like a network since elements of any other phase are comprised in each of the four phases. This complex network corresponds to the complexity of the change process;

• the change process becomes transparent to the people involved because the same steps show up everywhere within this process. Changes frequently affect people but they are necessary. They are more easily accepted if people understand the process and its sequences even if they do not yet know its outcome. It is similar to experiencing the changes of weather: humans are familiar with the change process without exactly knowing the outcome. This process-oriented understanding of change helps people to get used to the complexity of changes which are non-linear and show feedback loops and 'chaotic' patterns. These deeper insights may be seen as an essential learning outcome of change processes as described here.

These two different aspects of self-similarity - *structure* and *process* - can be integrated in the *management* of change processes in the following way: the whole

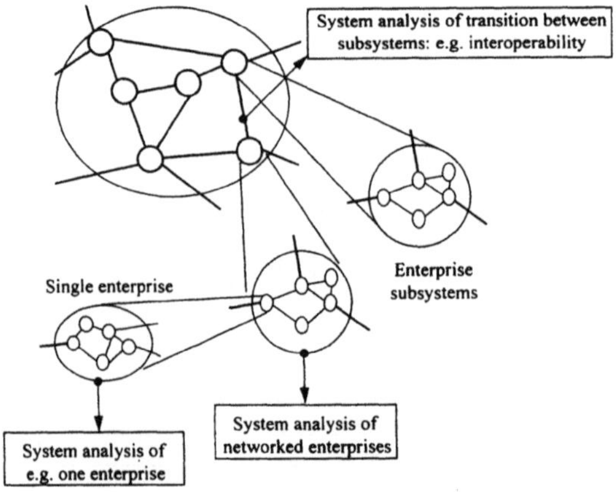

Figure 6. Self-similarity of systems.

enterprise is to be mirrored and re-produced in every one of the different teams which manage the change process of the enterprise. This strategy implies that not only the different departments of the enterprise are represented on the project teams (as in a "matrix" organisation) but all hierarchy levels of the company participate in each of the project teams.[19]

Systems Re-engineering

Applying these concepts in practice on enterprise change means to look at the organisation in a complementary way:

- on the one hand, to look at the technology-in-use and the data created by it, and to design new technological sub-systems to improve performance;
- on the other hand, to look at the communication and cooperation of all personnel involved in the business processes and workflows, and to draft new, improved organisational structures.

These two complementary views can be integrated in the *dual design approach*. The dual design approach is a set of principles to ensure appropriate development of both the technical and the human aspects of systems integrating humans *and* technology.[20]

Usually project engineers tend to head for fully automated concepts. This kind of approach - the technology based design - is represented by the left-hand triangle of Figure 7. Here the major part of design efforts, creativity and research is used to obtain a fully-automated system. Then, however, it is very difficult, if not impossible, to obtain jobs which take advantage of the strengths and competencies of people. Therefore, it is necessary to introduce a second approach, the working-process based design, in order to consider the human work situation as well. It raises the issue of how to solve the problem with a lower level of automation or computer use. This will result in a concept where tasks are mainly performed by people. The main part of design efforts, creativity and ideas will be put into this approach (see Fig. 7, right-hand triangle). Both the technology-based design and the working-process-based design should be used in parallel to obtain an optimum. Thus the dual design approach is complementary in the technology based design and in the working-process based design. The advantages and disadvantages of both concepts have to be compared and analysed. Such human-machine systems correspond to the demands of both the technical processes and the process of human work.

The development and implementation of new technologies usually lead to far-reaching changes of work in its context. In production industry or service enterprises, it is not considered important to adapt technology as much as possible to the existing work structures. This neglect of transition smoothness frequently leads to genuinely dangerous

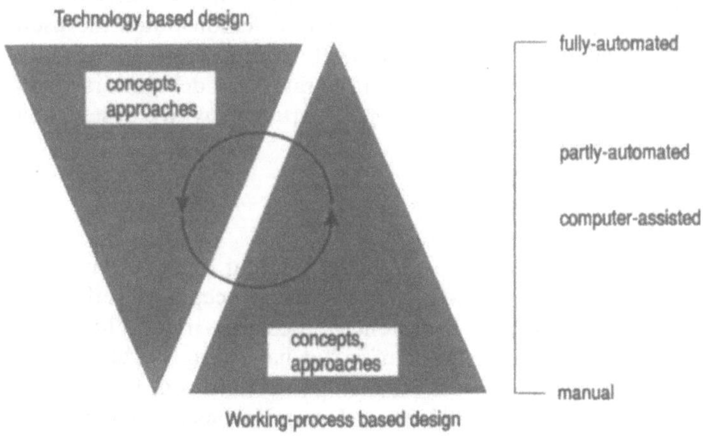

Figure 7. The dual design approach of designing human-machine systems.

developments in enterprises because software changes may paralyse the company for days (or even longer) before the new software system really works. Hence the re-engineering process needs to be structured by a series of distinctive steps:

- the step of *reverse engineering*: analysing and documenting the existing system, the tasks and needs of system users in situations of use, and the system context;
- the step of *forward engineering*: developing concepts of what the system is to do, and drafting this system. In a complex re-engineering project, it means to incrementally replace the old system by the new system as a basis for future developments. This process step may be called the 1:1 transition;
- the step of *system extension*: only now the new system can be extended to integrate and offer to the user, new functionalities which were the initial aim of the process.[21]

A Case Study

The concepts discussed above will be supported by experiences of a case study, where these design strategies were used.

For a big hospital (1500 beds), the diagnosis of the systems had shown that the official goals and strategies, oriented towards the hospitals' mission, were not reflected in the structure to a sufficient extent.

Thus the following changes were initiated by the HDZ/IMA in cooperation with administration and employees of the hospital:

- hospital department teams should operate on the basis of decentralised revenue and costs calculations;
- self-similar profit-centre structure were to be implemented on unit and ward level (see Fig. 8). The aim was to achieve effective teamwork between physicians, nurses and administration staff, who together are responsible for the whole ward and its costs;
- revenues and costs should be allocated over the "fractal" decentralised units; performance should be measured and accounted for, in a decentralised process;
- data requested from the central information system should be available to and used at the decentralised units (wards etc.)

Therefore, a decentralised EDP-structure was required, with user-friendly interfaces. It was considered useful to link such developments to the implementation of profit centre structures: the medical personnel of the individual departments should be held accountable for their own economic performance. However, accountability includes the creation of adequate knowledge on behalf of the personnel and quantifiable fee-back in order to enable objective-oriented economic performance.

Based on the development of self-similar profit centre structures, a flexible prototype was developed by the HDZ/IMA for decentralised, computer-assisted profit and cost accounting. Initial tests of the prototype in practice demonstrated the usefulness of this model for a detailed economic analysis, as well as its flexible applicability to different medical departments. These experiences illustrate that decentralised profit and cost accounting systems provide employees of the departments with a system enabling them to accept economic responsibilities.

Conclusions

The project team achieved both, the understanding of the complex system and its fundamental changes, by consciously applying the concepts of self-similarity. For the team, these concepts work like a system of beacons in un-chartered waters. The experiences of: "it repeats itself once again" – "we have seen that before, somewhere else, in a different context" - were the most staggering impressions of the team amidst apparent chaos of this huge, unwielding organisation. The project as a whole has proved that it is possible to understand and influence such organisations, after all.

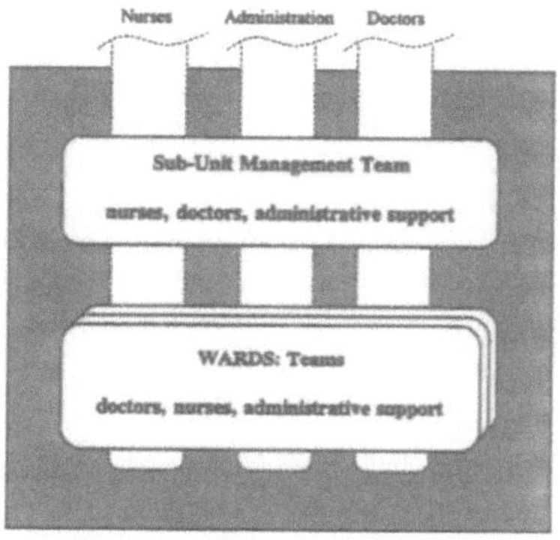

Figure 8. The hospital's new "Profit-Centre" structure. Hospital sub-unit (e.g. Neuro-Surgery), as an example (Frankfurt): about 24 sub-units as "Profit-Centres".

REFERENCES

1. K.J. Holyoak, Problem Solving, in: *Thinking. An Invitation to Cognitive Science*, D. N. Osherson and E.E. Smith eds., The MIT Press, Cambridge (1990).
2. R.C Lewontin., The Evolution of Cognition, in: *Thinking. An Invitation to Cognitive Science*, D. N. Osherson and E.E. Smith eds., The MIT Press, Cambridge (1990).
3. J Piaget, *L'équilibration des structures cognitives:\ problème central du développement*, Presses Universitaires de France, Paris (1975).
4. Newell, H.A. Simon, *Human problem solving*, Englewood Cliffs, NJ, Prentice-Hall (1972).
5. J.R. Anderson, *The architecture of cognition*, Harvard University Press (1983).
6. H.A. Simon, *The sciences of artificial*, MIT Press, Cambridge (1981).
7. W. Reitman, Heuristic decision procedures, open constraints, and the structure of ill-defined problems, in: *Human judgments and optimality*, W.Shelley and G.L. Bryan eds., Wiley, New York (1964).
8. J.P. Chaplin, T.S. Krawiec, *Systems and Theories of Phychology*, Holt, Rinehart and Winston (1968).
9. P. Thagard, *Mind. Introduction to Cognitive Science*, M.I.T. (1996)
10. P.N. Johnson-Laird and R.M.J. Byrne, *Deduction*, Lawrence Erlbaum Associates Ltd. (1991).
11. K. Duncker, On Problem Solving, *Psychological Monographs*, *58*, 270 (1945).
12. S. Russell and P. Norvig, *Artificial Intelligence – A Modern Approach*, Prentice-Hall, Upper Saddle River (1995).
13. R.E. Fikes and N.J. Nilsson, STRIPS: a new approach to the application of theorem proving to problem solving, *Artificial Intelligence*, 2 (1971).
14. R.A. Brooks, A robust layered control system for a mobile robot, *IEEE Journal of Robotics and Automation*, 2 (1986).
15. R.C. Arkin, *Behavior-Based Robotics*. M.I.T. Press, Boston (1998).
16. D.P. Hanna, *Designing Organisations for High Performance*, Reading, Massachusettes (1988).
17. H. Rieckmann and P.H. Weissengruber, Managing the unmanagable? - Oder: Lassen sich komplexe Systeme überhaupt noch steuern? - Offenes Systemmanagement mit dem OSTO-Systemansatz, in: *Management Development im Wandel*, H. Kraus, N. Kailer, K. Sandner eds., Wien (1990).
18. K. Henning and S. Marks, *Kommunikations- und Organisationsentwicklung*, Aachener Reihe Mensch und Technik, Aachen (1992).
19. G. Strina, Self-Similarity as a new principle for the design of innovation processes in production systems, *Proc. 5th IFAC Symposium Automated Systems Based on Human Skill*, 26.-28.9.1995, Berlin (1995).
20. B. Ochterbeck, *The Dual Design Approach applied to Rail-Road Container Terminals*, Düsseldorf: VDI (1989).
21. H. Kesselmeier, Software Reengineering of a touristic booking and information system in Europe. *Proc. 5th IFAC Symposium Automated Systems based on Human Skill*, 26.-28.09, Berlin (1995).

CREATIONS AND DISCOVERIES IN SCIENCE: THE ROLE OF ABDUCTIVE REASONING

Lorenzo Magnani

Computational Philosophy Laboratory
Department of Philosophy, University of Pavia
I 27100 Pavia – Italy
e-mail: lorenzo@philos.unipv.it

ABSTRACT

In different theoretical changes in theoretical systems we witness different kinds of discovery processes operating. Discovery methods are "data-driven", "explanation-driven" (abductive), and "coherence-driven" (formed to overwhelm contradictions). Sometimes there is a mixture of such methods: for example, an hypothesis devoted to overcome a contradiction is found by abduction. When a contradiction (or an anomaly) arises, consistency can be restored by rejecting or modifying any assumption which contributes to the derivation of contradiction. Hence, the derivation of inconsistency contributes to the search for alternative, and possibly new hypotheses.

The aim is to emphasise the significance of abduction in order to illustrate the problem solving process and to propose a unified epistemological model of scientific discovery. The model first describes the different meanings of the word abduction (creative, selective, to the best explanation, "model-based") in order to clarify their significance for epistemology and artificial intelligence. I will also illustrate abductive reasoning and its formal models in order to classify and analyse the different roles played by inconsistencies in different reasoning tasks. Many ways of "governing" inconsistencies are considered: from the methods activated in diagnostic settings and consistency-based models to the typical ones embedded in some forms of creative reasoning, from the interpretations in terms of conflicts and competitions to the actions performed on empirical and conceptual anomalies, from the question of generating inconsistencies by radical innovation to the connectionist treatment of coherence.

CHANGE IN THEORETICAL SYSTEMS

In different theoretical changes we witness different kinds of discovery processes operating. Discovery methods are *data-driven* (generalisations from observation and from experiments), *explanation-driven* (abductive), and *coherence-driven* (formed to overwhelm

Human and Machine Perception 2: Emergence, Attention, and Creativity,
edited by Cantoni *et al.* Kluwer Academic / Plenum Publishers, New York, 1999.

contradictions).[1] Sometimes there is a mixture of such methods: for example, an hypothesis devoted to overcome a contradiction is found by abduction. Therefore, contradiction and its reconciliation play an important role in philosophy, in scientific theories and in all kinds of problem-solving. It is the driving force underlying change (thesis, antithesis and synthesis) in the Hegelian dialectic and the main tool for advancing knowledge (conjectures and refutations[2] and proofs and counter-examples[3]) in the Popperian philosophy of science and mathematics.

Following Quine's line of argument against the distinction between necessary and contingent truths,[4] when a contradiction arises, consistency can be restored by rejecting or modifying any assumption which contributes to the derivation of contradiction: no hypothesis is immune from possible alteration. Of course there are epistemological and pragmatic limitations: some hypotheses contribute to the derivation of useful consequences more often than others, and some participate more often in the derivation of contradictions than others. For example it might be useful to abandon, among the hypotheses which lead to contradiction, the one which contributes least to the derivation of useful consequences; if contradictions continue to exist and the assessed utility of the hypotheses changes, it may be necessary to backtrack, reinstate a previously abandoned hypothesis and abandon an alternative instead.

Hence, the derivation of inconsistency contributes to the search for alternative, and possibly new hypotheses: for each assumption which contributes to the derivation of a contradiction there exists at least one alternative new system obtained by abandoning or modifying the assumption.

The classical example of a theoretical system that is opposed by a contradiction is the case in which the report of an empirical observation or experiment contradicts a scientific theory. Whether it is more beneficial to reject the report or the statement of the theory depends on the whole effect on the theoretical system. It is also possible that many alternatives might lead to non-comparable, equally viable, but mutually incompatible, systems.*

As Lakatos argues, in a mature theory with a history of useful consequences, it is generally better to reject an anomalous conflicting report than it is to abandon the theory as a whole. The cases in which we have to abandon a whole theory are very rare: a theory may be considered as a complex information system in which there is a collection of cooperating individual statements some of which are useful and more firmly held than others; propositions that belong to the central core of a theory are more firmly held than those which are located closer to the border, where instead rival hypotheses may coexist as mutually incompatible alternatives. Accumulating reports of empirical observations can help in deciding in favour of one alternative over another.

We have to remember that even without restoring consistency, an inconsistent system can still produce useful information. Of course from the point of view of classical logic we are compelled to derive any conclusion from inconsistent premises, but in practice efficient proof procedures infer only "relevant" conclusions with varying degrees of accessibility, as stated by the criteria of non-classical *relevant entailment*.[7]

We may conclude by asserting that contradiction, far from damaging a system, helps to indicate regions in which it can be changed (and improved). It is always better to produce mistakes and then correct them than to make no progress at all. Contradiction has a preference for strong hypotheses which are more easily falsified than weak ones; and moreover, hard hypotheses may more easily weakened than weak ones, which prove difficult subsequently to strengthen. In Magnani[8] we will consider a kind of "weak" hypothesis that is hard to negate; we will also illustrate the ways for making it easy, by

*Thagard proposes a very interesting computational account of scientific controversies in terms of so-called *explanatory coherence*,[1] which improves on Lakatos' classic one[5,6] see also subsection "Generating Inconsistencies by Radical Innovation" above.

explaining the logical and computational notion of *negation as failure*. In the following two sections we will briefly review abductive reasoning, its formal models, and various ways of governing inconsistencies.

ABDUCTION: CREATIVE, SELECTIVE, TO THE BEST EXPLANATION, MODEL-BASED

Many reasoning conclusions that do not proceed in a deductive manner are *abductive*. For instance, if we see a broken horizontal glass on the floor* we might explain this fact by postulating the effect of wind shortly before: this is not certainly a deductive consequence of the glass being broken (a cat may well have been responsible for it). Hence, abduction is the process of inferring certain facts and/or laws that render some sentences plausible, that explain some phenomenon or observation; it is the process of reasoning in which explanatory hypotheses are formed and evaluated.[9] First, it is necessary to show the connections between abduction, induction, and deduction and to stress the significance of abduction to illustrate the problem solving process. I have developed with others[10,11] an epistemological model of medical reasoning, called the *Select and Test Model* (ST–MODEL, see Magnani,[12] Stefanelli and Ramoni[13]) which can be described in terms of the classical notions of abduction, deduction and induction: it describes the different roles played by such basic inference types in developing various kinds of medical reasoning (diagnosis, therapy planning, monitoring). Abduction is becoming an increasingly popular term in artificial intelligence[14-18,1] especially in the field of medical knowledge-based systems.[19-21,12,11] A hundred years ago, Peirce[22] interpreted abduction essentially as an inferential *creative process* of generating a new hypothesis. There are two main epistemological meanings of the word abduction: 1) abduction that only generates plausible hypotheses (*selective* or *creative*)** and 2) abduction considered as *inference to the best explanation*, that also evaluates hypotheses. All we can expect of our "selective" abduction, is that it tends to produce hypotheses that have some chance of turning out to be the best explanation. Selective abduction will always produce hypotheses that give at least

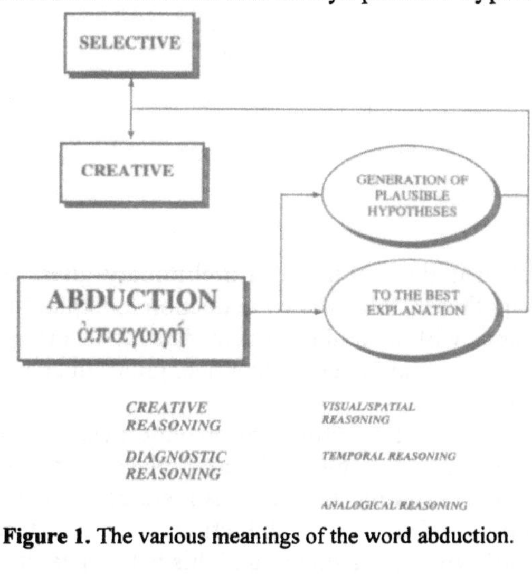

Figure 1. The various meanings of the word abduction.

*This event constitutes in its turn an *anomaly* that needs to be solved/explained.

**To illustrate from the field of medical knowledge, the discovery of a new disease and the manifestations it causes can be considered as the result of a creative abductive inference. Therefore, *creative* abduction deals with the whole field of the growth of scientific knowledge. This is irrelevant in medical diagnosis where instead the task is to *select* from an encyclopedia of pre-stored diagnostic entities.[11]

a partial explanation and therefore have a small amount of initial plausibility. In this respect abduction is more efficacious than the blind generation of hypotheses (see Fig. 1). In the view concerning abduction as inference to the best explanation advocated by Peirce one might require that the final chosen explanation be the *most* plausible.

Induction in its widest sense is an ampliative process of the generalisation of knowledge. Peirce distinguished three types of induction and the first was further divided into three sub-types. A common feature of all kinds of induction is the ability to compare individual statements: using induction it is possible to synthesise individual statements into general laws (types I and II), but it is also possible to confirm or discount hypotheses (type III). Clearly I am referring here to the latter type of induction, that in my model is used as the process of reducing the uncertainty of established hypotheses by comparing their consequences with observed facts.

Deduction is an inference that refers to a logical implication. Deduction may be distinguished from abduction and induction on the grounds that only in deduction is the truth of inference guaranteed by the truth of the premises on which it is based. All these distinctions need to be exemplified. To describe how the three inferences operate, it is useful to start with a very simple example dealing with diagnostic reasoning:

1. if a patient is affected by a beta thalassemia, his/her level of hemoglobin A2 is increased;
2. John is affected by a beta thalassemia;
3. John's level of hemoglobin A2 is increased.

By deduction we can infer (3) from (1) and (2); by induction we can go from a finite set of facts, like (2) and (3), to a universally quantified generalisation, like the piece of hematologic knowledge represented by (1). Starting from knowing - selecting - (1) and observing (3) we can infer (2) by performing a selective abduction. Such an inference is not affected by uncertainty, since the manifestation (3) is pathognomonic for beta-thalassemia. This is a special case, where there is not abduction because there is not "selection", in general clinicians very often have to deal with manifestations which can be explained by different diagnostic hypotheses: the abductive inference rule corresponds to the well-known fallacy called affirming the consequent:

$$\begin{array}{c} \psi \rightarrow \varphi \\ \varphi \\ \hline \psi \end{array}$$

Thus, *selective abduction* is the making of a preliminary guess that introduces a set of plausible diagnostic hypotheses, followed by deduction to explore their consequences, and by induction to test them with available patient data, i) to increase the likelihood of a hypothesis by noting evidence explained by that one, rather than by competing hypotheses, or ii) to refute all but one.

If during this first cycle new information emerges, hypotheses not previously considered can be suggested and a new cycle takes place: in this case the *nonmonotonic* character of abductive reasoning is clear. As stated above, there are two main epistemological meanings of the word abduction:[1] i) abduction that only generates plausible hypotheses (*selective* or *creative*) - and this is the meaning of abduction accepted in my epistemological model - and ii) abduction considered as *inference to the best explanation*, that also evaluates hypotheses. In the latter sense the classical meaning of abduction as *inference to the best explanation* (for instance in medicine, to the best diagnosis) is described in my epistemological model by the complete abduction - deduction - induction cycle. All we can expect of my "selective" abduction, is that it tends to produce hypotheses that have some chance of turning out to be the best explanation. Selective abduction will always produce hypotheses that give at least a partial explanation and therefore have a small amount of initial plausibility. In this respect abduction is more efficacious than the blind generation of hypotheses.

My epistemological model should be regarded as a very simple and schematic illustration of scientific theory change. In this case selective abduction is replaced by creative abduction and there exists a set of competing theories instead of diagnostic hypotheses. Furthermore the language of background scientific knowledge should be regarded as open: in the case of competing theories, as they are studied using the epistemology of theory change, we cannot - contrary to Popper's viewpoint[2] - reject a theory simply because it fails occasionally. If for example such a theory is simpler and explains more significant data than its competitors, then it can be accepted as the best explanation (see below "Generating Inconsistencies by Radical Innovation").

We should remember, as Peirce noted, that abduction plays a role even in relatively simple *visual phenomena*. *Visual abduction*, a special form of abduction, occurs when hypotheses are instantly derived from a stored series of previous similar experiences. It covers a mental procedure that tapers into a non-inferential one, and falls into the category called "perception". Philosophically, *perception* is viewed by Peirce as a fast and uncontrolled knowledge-production procedure. Perception, in fact, is a vehicle for the instantaneous retrieval of knowledge that was previously structured in our mind through inferential processes. By perception, knowledge constructions are so instantly reorganised that they become habitual and diffuse and do not need any further testing. Many visual stimuli are ambiguous, yet people are adept at imposing order on them: "We readily form such hypotheses as that an obscurely seen face belongs to a friend of ours, because we can thereby explain what has been observed".[18,p.53] This kind of image-based hypothesis formation can be considered as a form of *visual abduction*[23] (see "Finding Inconsistencies: Empirical and Conceptual Anomalies" below).

We have to say that visual and analogical reasoning are productive in scientific concept formation too; scientific concepts do not pop out of heads, but are elaborated in a problem solving process that involves the application of various procedures: this process is a *reasoned process*. We know that scientific concept formation has been ignored because of the accepted view that no "logic of discovery" - either deductive, inductive, or abductive algorithms for generating scientific knowledge - is possible. The methods of discovery involve use of *heuristic* procedures: cognitive psychology, artificial intelligence, and computational philosophy have established that heuristic procedures are reasoned. Analogical reasoning is one such problem solving procedure, and some reasoning from imagery is a form of analogical reasoning.[24]

How does this kind of analogical and/or imagery reasoning function in problem solving? Nersessian[25,26] has demonstrated that history of science abounds with instances of the use of imagery and of analogy to transform vague notions into scientifically viable conceptualisations of a domain. Her analysis deals with the important case of the use of imagery and analogy by Faraday and Maxwell in the costruction of the concept of field. The concept of field had its origins in vague speculations about processes in the regions surrounding bodies and charges that might contribute to their action upon one another. In articulating a field representation for electric and magnetic actions, Faraday used primarily qualitative concepts and reasoned from imagery figures. He created a field representation for electric and magnetic actions by reasoning from an imagery representation of the "lines of force" that are formed when iron filings are sprinkled around a magnetic source. Many features of "lines" are incorporated into his field concept. He discussed many actions as "expanding", "bending", "being cut". All the forces of nature are unified and interconvertible through various motions of the lines of forces, and matter, itself, is nothing but point centers of converging lines of force. This representation enabled Faraday to express a quantitative relationship between the number of lines cut and the intensity of the induced force. At the end of this research Faraday introduced a pictorial representation that was to play an important role in Maxwell's construction of the quantitative field concept. Use of analogy and imagery in ordinary and scientific problem solving is very complex. Nevertheless, we may observe in many cases all the features of a *productive, creative* mapping, where such "transfer of knowledge" is essential to the development of a new concept. Imagery representations appear to function analogically. The value of an imagery representation is that it makes some structural relations immediately evident.

Visual abduction, but also many kinds of abductions involving analogies, diagrams, thought experimenting, etc., can be called *model-based.**

GOVERNING INCONSISTENCIES IN ABDUCTIVE REASONING

Formal Models of Abductive and Consistency-Based Reasoning

Many attempts have been made to model abduction by developing some formal tools in order to illustrate its computational properties and the relationships with the different forms of deductive reasoning.[29-34] Some of these formal models of abductive reasoning, for instance,[35] are based on the theory of the *epistemic state* of an agent,[36,37] where the epistemic state of an individual is modeled as a consistent set of beliefs that can change by expansion and contraction.** We shall discuss the nature of the kinds of inconsistencies captured by these formalisms and show how they do not adequately account for some roles played by anomalies, conflicts, and contradictions in many forms of explanatory reasoning.

Deductive models of abduction may be characterised as follows (see Fig. 2): an explanation for β relative to background theory T will be any α that, together with T, entails β (normally with the additional condition that $\{\alpha\} \cup T$ be consistent). Such theories are usually generalised in many directions: first of all by showing that explanations entail their conclusions only in a *defeasible* way (there are many potential explanations), so joining the whole area of the so-called nonmonotonic logic or of the probabilistic treatments; second, trying to show how some of the explanations are relatively implausible, elaborating suitable technical tools (for example in terms of modal logic) able to capture the notion of preference among explanations. Hence, we may require that an explanation

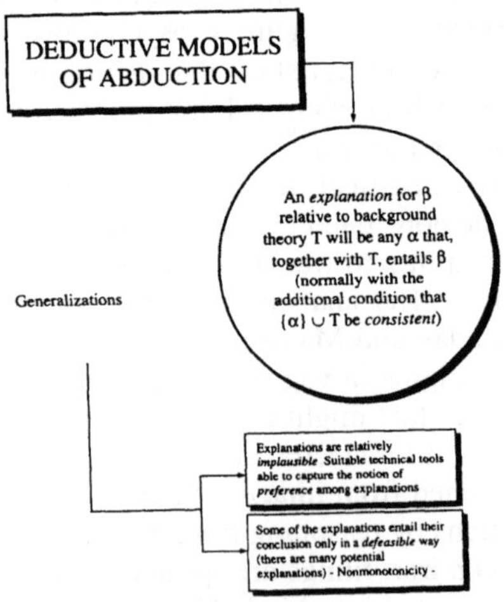

Figure 2. Deductive models of abduction.

*Model-based reasoning and the so-called *constructive modeling* and *generic modeling* in scientific discovery are illustrated by Nersessian.[27,28]

**Levi's theory of suppositional reasoning is also related to the problem of "belief change".[38]

makes the observation simply sufficiently probable[39] or that the explanations that are more likely will be the "preferred" explanations: the involvement of a cat in breaking the glass is less probable than the effect of wind. Finally, the deductive model of abduction does not authorise us to explain facts that are inconsistent with the background theory notwithstanding the fact that these explanations are very important and ubiquitous, for instance in diagnostic applications, where the facts to be explained contradict the expectation that the system involved is working according to specification.

Boutilier and Becher[35] provide a formal account of the whole question in term of belief revision: if believing A is sufficient to induce belief in B, then A (epistemically) *explains* B; the situation can be semantically illustrated in terms of an ordering of plausibility or normality which is able to represent the epistemic state of an agent. The conflicting observations will require explanations that compel the agent to withdraw its beliefs (hypotheses), and the derived conditional logic is able to account for explanations of facts that *conflict* with the existing beliefs. The authors are able to reconstruct, within their framework, the two main paradigms of model-based diagnosis, *abductive*,[40,41] and *consistency-based*,[42,32] providing an alternative semantics for both in terms of a plausibility ordering over possible worlds.

Let us resume the kinds of change considered in the original belief revision framework (see Fig. 3). The *expansion* of a set of beliefs K taken from some underlying language (considered to be the closure of some finite set of premise KB, or *knowledge base*, so $K=Cn(KB)$) by a piece of new information A is the belief set $K + A = Cn(K \cup AB)$. The addition happens "regardless" of whether the larger set is *consistent*. The case of *revision* happens when $K \models \neg A$, that is when the new A is *inconsistent* with K and we want to maintain consistency: some beliefs in K must be withdrawn before A can be accommodated: $K - A$. The problem is that it is difficult to detect which part of K has to be withdrawn. The least "entrenched" beliefs in K should be withdrawn and A added to the "contracted" set of beliefs. The loss of information has to be as small as possible so that "no belief is given up unnecessarily".[37] Hence, *inconsistency resolution* in belief revision framework is captured by the concept of revision. Another way of belief change is the process of *contraction*. When a belief set K is contracted by A, the resulting belief set $K + A$ is such that A is no longer held, without adding any new fact.

After having explained the distinction between predictive explanations and "might" explanations, that merely allow an observation, and do not predict it, Boutilier and Becher show in the cited article how model-based diagnoses can be accounted for in terms of their

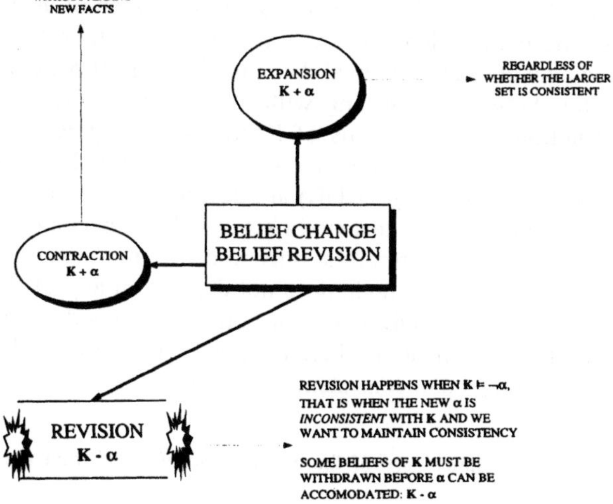

Figure 3. Belief revision.

new formal model of belief revision.

The *abductive* model-based reasoning*[40,41,43] illustrated by some models, such as Poole's Theorist, allows many possible explanations, weak and predictive (so presenting a paraconsistent behaviour: a non-predictive hypothesis can explain both a proposition and its negation). This old model, embedded in the new formal framework, acquires the possibility of discriminating certain explanations as preferred to others. Reiter's *consistency-based* diagnosis is devoted[32] to ascertain why a correctly designed system is not working according to its features. Because certain components may fail, the system description also contains some abnormality predicates (the absence of them will render the description inconsistent with an observation of an incorrect behaviour). The consistency-based diagnosis concerns any set of components whose abnormality makes the observation consistent with the description of the system. A principle of parsimony is also introduced to capture the idea of preferred explanations/diagnoses. Since the presence of fault models renders Reiter's framework incorrect, new more complicated notions are introduced in de Kleer et al.,[42] where the presence of a complete fault model ensures that predictive explanations may be given for "every" abnormal observation. Without any description of correct behaviour any observation is consistent with the assumption that the system works correctly. Hence, a complete model of correct behaviour is necessary if we want the consistency-based diagnosis to be useful. The idea of consistency that underlies this kind of diagnostic reasoning is the following: any inconsistency (anomalous observation) is an aberrant behaviour that can usually be accounted for by finding some set of components of a system that, if behaving abnormally, will entail or justify the actual observation.

Without doubt the solution given by Boutilier and Becher furnishes a more satisfying qualitative account of the choice among competing explanations than Gärdenfors'[44] in terms of "epistemic entrenchment",** which tries to capture the idea of an ordering of beliefs according to our willingness to withdraw them when necessary. Moreover, the new formal account in terms of belief revision is very powerful in shedding new light on the old model-based accounts of diagnostic reasoning.

The framework of belief revision is sometimes called *coherence approach*.[45] In this approach it is important that the agent holds some beliefs just as long as they are consistent with the agent's remaining beliefs. Inconsistent beliefs do not describe any world, and so are unproductive; moreover, the changes must be epistemologically conservative in the sense that the agent maintains as many of its beliefs as possible when it adjusts its beliefs to the new information. It is contrasted to the *foundations approach*, according to which beliefs change as the agent adopts or abandons satisfactory reasons (or justifications). This approach is exemplified by the well-known "Reason Maintenance Systems" (RMS) or "Truth Maintenance Systems" (TMS),[45] elaborated in the area of artificial intelligence to cooperate with an external problem solver. In this approach the role of inconsistencies is concentrated on the negations able to invalidate justifications of beliefs; moreover, as there are many similarities between reasoning with incomplete information and acting with inconsistent information, the operations of RMS concerning revision directly involve logical consistency, seeking to solve a conflict among beliefs. The operations of *Dependency-Directed Backtracking* (DDB) are devoted to this aim: RMS informs DDB whenever a contradiction node (for instance a set of beliefs) becomes believed, then DDB attempts to remove reasons and premises, only to defeat nonmonotonic assumptions: "If the argument for the contradiction node does not depend on any of these (i.e., it consists entirely of monotonic reasons), DDB leaves the contradiction node in place as a continuing belief",[46,p.36] so leaving the conflicting beliefs intact if they do not depend on defeasible assumptions, and presenting a paraconsistent behaviour.

*Please distinguish here the technical use of the attribute *model-based* from the epistemological one I introduced in the previous section.

**Which of course may change over time or with the state of belief.

Both in the coherence and foundations approach the changes of state have to be *epistemologically conservative*: as already said above the agent maintains as many of its beliefs as possible when it adjusts its beliefs to the new information, thus following Quine's idea of "minimum mutilation".[47] We have now to notice some limitations of the formal models in accounting for other kinds of inconsistencies embedded in many reasoning tasks.

More Conflictual Features, Creative Settings, Coherence

If we want to deal with the nomological and most interesting creative aspects of abduction we are first of all compelled to consider the whole field of the growth of scientific knowledge cited above. We have anticipated that abduction has to be an inference permitting the derivations of *new* hypotheses and beliefs. Some explanations consist of certain facts (initial conditions) and universal generalisations (that is scientific laws) that deductively entail a given fact (observation), as showed by Hempel in his *law covering model* of scientific explanation.[48] If T is a theory illustrating the background knowledge (a scientific or common sense *theory*) the sentence α explains the fact (observation) β just when$\{\alpha\} \cup T \models \beta$; it is difficult to govern the question involving nomological and causal aspects of abduction and explanation in the framework of the belief revision illustrated in the previous section: we would have to deal with a kind of belief revision that permits us to alter a theory with new conditionals.

We may also see belief change from the point of view of *conceptual change*, considering concepts either cognitively, like mental structures analogous to data structures in computers, or, epistemologically, like abstractions or representations that presuppose questions of justification. Belief revision - even if extended by formal accounts such as illustrated above* - is able to represent cases of conceptual change such as adding a new instance, adding a new weak rule, adding a new strong rule (see Thagard,[1,pp.34-39] for details), that is, cases of addition and deletion of beliefs, but fails to take into account cases such as adding a new part-relation, adding a new kind-relation, adding a new concept, collapsing part of a kind-hierarchy, reorganising hierarchies by branch jumping and tree switching, in which there are reorganisations of concepts or redefinitions of the nature of a hierarchy. These last cases are the most evident changes occurring in many kinds of creative reasoning, for example in science. Related to some of these types of conceptual change are different varieties of inconsistencies (see Fig. 4).

Finding Inconsistencies: Empirical and Conceptual Anomalies. It may be said that logical accounts of abduction described above certainly illustrate much of what is

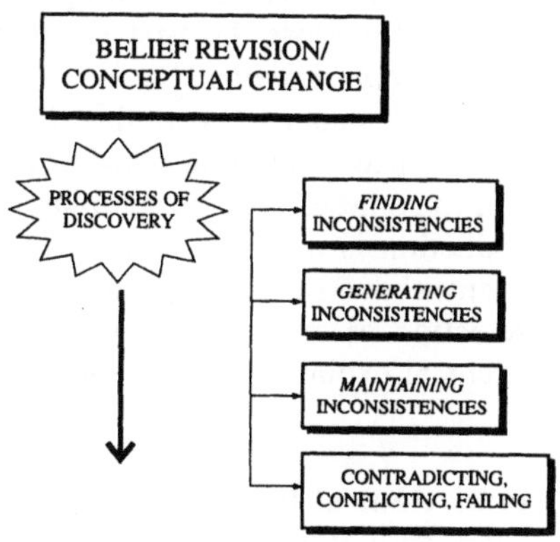

Figure 4. Inconsistencies and discovery.

*Or developed by others, see for example, Katsuno and Mendelzon[49] and Cross and Thomason[50].

important in abductive reasoning, especially the objective of selecting a set of hypotheses (diagnoses, causes) that are able to dispense good (preferred) explanations of data (observations), but fail in accounting for many cases of explanations occurring in science or in everyday reasoning. For example they do not capture: i) the role of statistical explanations, where what is explained follows only probabilistically and not deductively from the laws and other tools that do the explaining; ii) the sufficient conditions for explanation; iii) the fact that sometimes the explanations consist of the application of schemas that fit a phenomenon into a pattern without realising a deductive inference; iv) the idea of the existence of high-level kinds of *creative* abductions I cited above; v) the existence of model-based abductions (for instance visual and diagrammatic); vi) the fact that explanations usually are not complete but only furnish *partial* accounts of the pertinent evidence (see Thagard and Shelley[9]).

Moreover, the logical accounts of abduction certainly elucidate many kinds of inconsistency government, which nevertheless reduce to the act of finding contradictions able to generate the withdrawal of some hypotheses, beliefs, reasons, etc.: these contradictions always emerge at the level of data (observations), and consistency is restored at the theoretical level.* This view may distract from important aspects of other kinds of reasoning that involve intelligent abductive performances.

For example, *empirical anomalies* are not alone in generating impasses, there are also the so-called conceptual anomalies. In science, first and foremost, empirical anomaly resolution involves the localisation of the problem at hand within one or more constituents of the theory. It is then necessary to produce one or more new hypotheses to account for the anomaly, and finally, these hypotheses need to be evaluated so as to establish which one best satisfies the criteria for theory justification. Hence, anomalies require a change in the theory, yet once the change is successfully made, anomalies are no longer anomalous but in fact are now resolved. General strategies for anomaly resolution, as well as for producing new ideas and for assessing theories, have been studied by Darden.[51]

The so-called *conceptual problems* represent a particular form of anomaly. In addition, resolving conceptual problems may involve satisfactorily answering questions about the nature of theoretical entities. Nevertheless such conceptual problems do not arise directly from data, but from the nature of the claims in the principles or in the hypotheses of the theory. It is far from simple to identify a conceptual problem that requires a resolution, since, for example, a conceptual problem concerns the adequacy or the ambiguity of a theory, and yet also its incompleteness or (lack of) evidence. In Magnani[52] I present some examples derived from the historical discovery of non-Euclidean geometries which illustrate the relationships between strategies for anomaly resolution and explanatory and productive visual thinking: I consider how visual thinking is relevant to hypothesis formation and scientific discovery and explore the first epistemological and cognitive features of what I described above as *visual abduction*.

The fact that inconsistencies may occur also at the theoretical level is further emphasised if we consider that in science or in legal reasoning,[1] hypotheses are mainly *layered*. Hence, the organisation of hypotheses is more complex than the one illustrated in previous formal models, and abduction is not only a matter of mapping from sets of hypotheses to a set of data. In many abductive settings there are hypotheses that explain other hypotheses so that the selection or creation of explanations is related to these relationships. This kind of hierarchical explanations has also been studied in the area of probabilistic belief revision.[38]

Generating Inconsistencies by Radical Innovation. The case of conceptual change such as adding a new part-relation, adding a new kind-relation, adding a new concept, collapsing part of a kind-hierarchy, reorganising hierarchies by branch jumping and tree

*We have to remember that the logical models in some cases exhibit a sort of paraconsistent behaviour.

switching, in which there are reorganisations of concepts or redefinitions of the nature of a hierarchy are the most evident changes occurring in many kinds of *creative reasoning*, for instance in the growth of scientific knowledge.

When a scientist introduces a new hypothesis, especially in the field of natural sciences, he is interested in the potential rejection of an old theory or of an old knowledge domain. Consistency requirements we described in the framework of deductive models, governing hypothesis withdrawal in various ways, would arrest further developments of the new abduced hypothesis. In the scientist's case there is not the deletion of the old concepts, but rather the *coexistence* of two rival and competing views. Consequently we have to consider this competition as a form of epistemological, and non logical inconsistency. For instance two scientific theories are conflicting because they compete in explaining shared evidence.

The problem has been studied in Bayesian terms but also in connectionist ones, using the so-called theory of explanatory coherence,[1] which deals with the epistemological (but sometimes pragmatical) reasons for accepting a whole set of explanatory hypotheses conflicting with another one. In some cognitive settings, such as the task of comparing a set of hypotheses and beliefs incorporated in a scientific theory with the one of a competing theory, we have to consider a very complex set of criteria (to ascertain which composes the best explanation), that goes beyond the mere simplicity or explanatory power. The minimality criteria included in some of the formal accounts of abduction, or the idea of the choice among preferred models cited above, are not sufficient to illustrate more complicated cognitive situations.

Maintaining Inconsistencies. As noted in the previous subsection, when we create or produce a new concept or belief that competes with another one, we are compelled to maintain the derived inconsistency until the possibility of rejecting one of the two becomes feasible. Other cognitive and epistemological situations present a sort of paraconsistent behaviour: a typical kind of *consistency maintenance* is the well-known case of scientific theories that face anomalies. As noted above, explanations are usually not complete but only furnish partial accounts of the pertinent evidence: not everything has to be explained.

Newtonian mechanics is forced to cohabit with the anomaly of perihelion of Mercury until the development of the theory of relativity, but it also has to stay with its false prediction about the motion of Uranus. In diagnostic reasoning too, it is necessary to make a diagnosis even if many symptoms are not explained or remain mysterious. In this situation we again find the similarity between reasoning in the presence of inconsistencies and reasoning with incomplete information already stressed. Sometimes scientists may generate the so-called *auxiliary hypotheses*,[5] justified by the necessity of overcoming these kinds of inconsistencies: it is well-known that the auxiliary hypotheses are more acceptable if able to predict or explain something new (the making of the hypothesis of the existence of another planet, Neptune, was a successful way - not an *ad hoc* manoeuvre - of eliminating the anomaly of the cited false prediction).

Contradicting, Conflicting, Failing. Considering the *coherence* of a conceptual system as a matter of the simultaneous satisfaction of a set of positive and negative constraints leads to the *connectionist models* (also in computational terms) of coherence. In this light logical inconsistency becomes a relation that furnishes a *negative* constraint and entailment becomes a relation that provides a *positive* constraint. For example, as already noted, some hypotheses are inconsistent when they simply compete, when there are some pragmatic incompatibility relations, when there are incompatible ways of combining images, etc.[53]

From the viewpoint of the connectionist model of coherence, it spontaneously allows the situations in which there is a set of accepted concepts containing an inconsistency (see previous subsection), for example in the case of anomalies: the system at hand may at any rate have a maximised coherence, when compared to another system. Moreover, "another interesting case is the relation between quantum theory and general relativity, two theories which individually possess enormous explanatory coherence. According to the eminent mathematical physicist Edward Witten, "the basic problem in modern physics is that these two pillars are incompatible [...]". Quantum theory and general relativity may be incompatible, but it would be folly given their independent evidential support to suppose that one must be rejected" (*ibid.*).

REFERENCES

1. P. Thagard, *Conceptual Revolutions*, Princeton University Press, Princeton (1992).
2. K. Popper, *Conjectures and Refutations. The Growth of Scientific Knowledge*, Routledge and Kegan Paul, London (1963).
3. I. Lakatos, *Proofs and Refutations. The Logic of Mathematical Discovery*, Cambridge University Press, Cambridge (1976).
4. W.V.O. Quine, Two dogmas of empiricism, *Philosophical Review*, 40, pp. 113-127 (1951). Also in: *From a Logical Point of View*, W.V.O. Quine, Hutchinson, London, pp. 20-46 (1953) (1961 2nd).
5. I. Lakatos, Falsification and the methodology of scientific research programs, in: *Criticism and the Growth of Knowledge*, I. Lakatos and A. Musgrave eds., Cambridge University Press, Cambridge, pp. 91-195 (1970).
6. I. Lakatos, History of science and its rational reconstructions, in: *PSA 1970: In memory of Rudolf Carnap*, R. Buck and R. S. Cohen eds., Reidel, Dordrecht (1971).
7. A. Anderson and N. Belnap, *Entailment*, Princeton University Press, Princeton (1975).
8. L. Magnani, Withdrawing hypotheses by negation as failure, in: *Essays in Honor of Imre Toth*, K. Chemla and S. Probst eds., forthcoming.
9. P. Thagard and C.P. Shelley, Abductive reasoning: logic, visual thinking, and coherence, in: *Logic and Scientific Methods*, M.L. Dalla Chiara, K. Doets, D. Mundici, and J. van Benthem eds., Kluwer, Dordrecht, pp. 413-427 (1997).
10. G. Lanzola, M. Stefanelli, G. Barosi, and L. Magnani, NEOANEMIA: A knowledge-based system emulating diagnostic reasoning, *Computers and Biomedical Research*, 23, pp. 560-582 (1990).
11. M. Ramoni, M. Stefanelli, L. Magnani, and G. Barosi, An epistemological framework for medical knowledge-based systems, *IEEE Transactions on Systems, Man, and Cybernetics*, 22(6), pp. 1361-1375 (1992).
12. L. Magnani, Abductive reasoning: philosophical and educational perspectives in medicine, in: *Advanced models of cognition for medical training and practice*, D.A. Evans and V.L. Patel eds., Springer, Berlin, pp. 21-41 (1992).
13. M. Stefanelli and M. Ramoni, Epistemological constraints on medical knowledge-based systems, in: *Advanced Models of Cognition for Medical Training and Practice*, D.A. Evans and V.L. Patel eds., Springer, Berlin, pp. 3-20 (1992).
14. I. Peng and I.A. Reggia, A probabilistic causal model for diagnostic problem solving I: integrating symbolic causal inference with numeric probabilistic inference, *IEEE Transactions on Systems, Man, and Cybernetics*, 17, pp. 146-162 (1987).
15. I. Peng and I.A. Reggia, A probabilistic causal model for diagnostic problem solving II: diagnostic strategy, *IEEE Transactions on Systems, Man, and Cybernetics*, 17, pp. 395-406 (1987).
16. H.E. Pople, On the mechanization of abductive logic, in: *Proceedings of the International Joint Conference on Artificial Intelligence*, 8, pp. 147-152 (1973).
17. J.A. Reggia, S.N. Dana, and Y.W. Pearl, Expert systems based on set covering model, *International Journal on Man-Machine Studies*, 19, pp. 443-460 (1983).
18. P. Thagard, *Computational Philosophy of Science*, The MIT Press, Cambridge, MA (1988).
19. J.R. Josephson, B. Chandrasekaran, J.W.jr Smith, and M.C. Tanner, Abduction by classification and assembly, in *PSA 1986*, vol. 1, Philosophy of Science Association, pp. 458-470 (1986).
20. J.R. Josephson and S.G. Josephson, *Abductive Inference. Computation, Philosophy, Technology*, Cambridge University Press, Cambridge (1994).
21. L. Magnani, Epistémologie de l'invention scientifique, *Communication & Cognition*, 21, 3/4, pp. 273-291 (1988).
22. C.S. Peirce, *Collected Papers*, 8 vols., C. Harstone, P. Weiss, and A. Burks eds., Harvard University Press, Camdridge, MA, pp. 1931-1958.
23. L. Magnani, S. Civita, and G. Previde Massara, Visual cognition and cognitive modeling, in: *Human and Machine Vision: Analogies and Divergences*, V. Cantoni ed., New York, Plenum Press, pp. 229-243 (1994).

24. H.J. Holyoak and P. Thagard, *Mental Leaps. Analogy in Creative Thought*, The MIT Press, Cambridge, MA.
25. N.J. Nersessian, Reasoning from imagery and analogy in scientific concept formation, in: *PSA 1988*, vol 1, East Lansing: Philosophy of Science Association, A. Fine and J. Leplin eds., pp. 41-47 (1988).
26. N.J. Nersessian, Opening the black box: cognitive science and history of science. Technical Report GIT-COGSCI 94/23, July. Cognitive Science Report Series. Atlanta: Georgia Institute of Technology. Partially published in *Osiris* 10, pp. 194-211 (1995).
27. N. Nersessian, Constructive Modeling in Creating Scientific Understanding, *Science & Education*, 4, pp. 203-226 (1995).
28. N. Nersessian, T.W. Griffith, and A. Goel, Constructive Modeling in Scientific Discovery, Technical Report, Georgia Institute of Technology, Atlanta, GA (1997).
29. T. Bylander, D. Allemang, M.C. Tanner, and J.R Josephson, The computational complexity of abduction, *Artificial Intelligence*, 49, pp. 25-60 (1991).
30. K. Konolige, Abduction versus closure in causal theories, *Artificial Intelligence*, 53, pp. 255-272 (1992).
31. H.J. Levesque, A knowledge level account of abduction, in: *Proceedings IJCAI-89*, Dedroit, MI, pp. 1061-1067 (1989).
32. R. Reiter, A theory of diagnosis from first principles, *Artificial Intelligence*, 32, pp. 57-95 (1987).
33. M. Shanahan, Prediction is deduction but explanation is abduction, in: *Proceedings IJCAI-89*, Detroit, MI, pp. 1140-1145 (1989).
34. R. Reiter and J. De Kleer, Foundations of assumption-based truth maintenance systems: preliminary report, in: *Proceedings AAAI-87*, Seattle, WA, pp. 183-188 (1987).
35. C. Boutilier and V. Becher, Abduction as belief revision, *Artificial intelligence*, 77, pp. 43-94 (1995).
36. C. Alchourrón, P. Gärdenfors, and P. Makinson, On the theory of logic change: partial meet functions for contractions and revision, *Journal of Symbolic Logic*, 50, pp. 510-530 (1985).
37. P. Gärdenfors, *Knowledge in Flux*, The MIT Press, Cambridge (1988).
38. I. Levi, *For the Sake of the Argument. Ramsey Test Conditionals, Inductive Inference, and Nonmonotonic Reasoning*, Cambridge University Press, Cambridge (1996).
39. J. Pearl, *Probabilistic Reasoning in Intelligent Systems*, Morgan Kaufmann, San Mateo, CA (1988).
40. D. Poole, A logical framework for default reasoning, *Artificial Intelligence*, 36, pp. 27-47 (1988).
41. D. Poole, Representing diagnostic knowledge for probabilistic Horn abduction, in: *Proceedings IJCAI-91*, Sydney, NSW, pp. 1129-1135 (1991).
42. J. de Kleer, A.K. Mackworth, and R. Reiter, Characterizing diagnoses, in: *Proceedings AAAI-90*, Boston, MA, pp. 324-330 (1990).
43. G. Brewka, Preferred subtheories: an extended logical framework for default reasoning, *Proceedings IJCAI-89*, Detroit, MI, pp. 1043-1048 (1989).
44. P. Gärdenfors ed., *Belief Revision*, Cambridge University Press, Cambridge (1992).
45. J. Doyle, A truth maintenance system, *Artificial Intelligence*, 12, pp. 231-272 (1979).
46. J.Doyle, Reason maintenance and belief revision: foundations versus coherence theories, in: *Belief Revision*, P. Gärdenfors ed., Cambridge University Press, Cambridge, pp. 29-51 (1992).
47. W.V.O. Quine, *Philosophy of Logic*, Prentice-Hall, Englewood Cliffs, NJ (1979).
48. C.G. Hempel, *Philosophy of Natural Science*, Prentice-Hall, Englewood Cliffs, NJ (1966).
49. H. Katsuno and A. Mendelzon, On the difference between updating a knowledge base and revising it, in: *Belief Revision*, P. Gärdenfors ed., Cambridge University Press, Cambridge, pp. 183-203 (1992).
50. C. Cross and R.H. Thomason, Conditionals and knowledge-base update, in: *Belief Revision*, P. Gärdenfors ed., Cambridge University Press, Cambridge, pp. 247-275 (1992).
51. L. Darden, *Theory Change in Science: Strategies from Mendelian Genetics*, Oxford University Press, Oxford (1991).
52. L. Magnani, *Ingegnerie della conoscenza. Introduzione alla filosofia computazionale*, Marcos y Marcos, Milan (1997).
53. P. Thagard and K. Verbeurgt, Coherence as constraint satisfaction, *Cognitive Science*, 22(1), pp. 1-24 (1998).

ANALOGY PERCEPTION AND CREATIVITY MECHANISMS

Bruno Vivicorsi

CREPCO UMR 6561
(Centre de Recherche en Psychologie Cognitive)
Université de Provence & CNRS
29, av. R. Schuman
13621 Aix-en-Provence Cedex 1 - France
e-mail: vivicors@newsup.univ-mrs.fr

INTRODUCTION: FROM PERCEPTION TO CREATIVITY

Numerous works in cognitive psychology show that perception can be oriented by stimulus features and knowledge. Perception is then not a real tracing. Nevertheless, intelligence, or situation adaptation, takes root in perception, in particular when we can perceive analogies between situations. According to Hofstadter, creativity results from variations around what is perceived as the same in situations. Consequently, mechanisms involved in the creativity are seen as analogy perception processes: a high-level perception process (perceiving a situation in a certain way) and a perception-conceptualisation loop (perception entails concepts activation that can orients perception, and so on). These processes permit to be leaded towards a specific resolution without determined solution. These mechanisms are generally used in activities that require mental fluidity and imply some consequences about concepts nature and cognitive architecture.

PERCEPTION IN COGNITIVE PSYCHOLOGY

Visual perception process has particularly drew careful psychological studies since more than 150 years. In favour of the visible no-error perception of the world we can exhibit, we tend to believe as Massaro[1,p.255] said that «we see the world in a certain way because that's the way the world is», but nothing of the sort. What is more, the perceiver is nor tape recorder nor photocopy machine. Two no exclusive processing are used: the data-driven process, i.e. guided by the stimuli features, and the concepts-driven process, i.e. guided by the knowledge background of the perceiver. The weight of each processing varies

Human and Machine Perception 2: Emergence, Attention, and Creativity,
edited by Cantoni *et al.* Kluwer Academic / Plenum Publishers, New York, 1999.

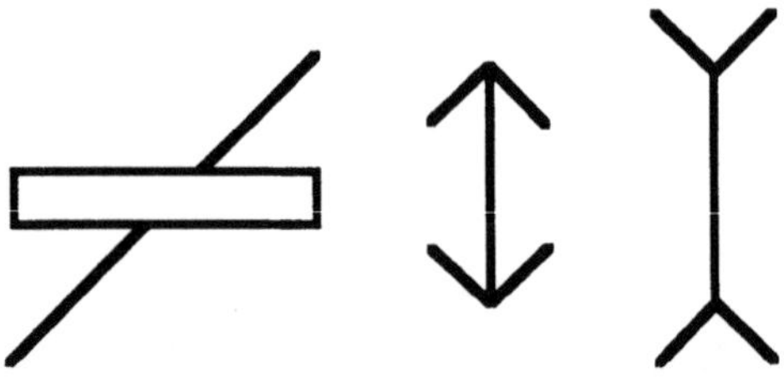

Figure 1. Two famous illusions showing the possible influence of stimulus features.

according to the situation we are in front of. "Visual illusions"[2] show us the way they can or not interact.*

From stimulus to cognitive system

We can bring the bottom-up (data-driven) process to the fore by using geometrical illusions (Figure 1). The first experimental one is the transparent rhomboid of L.A. Necker, in 1832, which spontaneously reverses in depth, presented and known today as the *Necker cube*. Johann Zöllner, Johann Poggendorf, in 1860, Ewald Hering in 1861, and Wilhelm Wundt in 1896 presented illusions on parallel lines which are perceived displaced or bowed. The double-headed-arrow figure, devised with 15 variations by Franz Müller-Lyer in 1889, shows deeply how is sometimes difficult to not be leaded by the stimulus feature: indeed, the central axis is perceived as longer with the wings directed outward than inward, and students don't believe me when I tell them that shafts are equal. It is the same thing for the railway lines illusion devised by Mario Ponzo in 1913: a horizontal line is perceived as longer when it is farther along the converging railroad tracks. A variation can be seen in the drawing «To make the tables turn» by Roger N. Shepard:[3,4] we have to measure dimensions of the tables to make us sure that tables have the same ones.

All these examples show that some features of the stimulus can make us to perceive it in a certain way, and illusions continue to persist even though we know that what we are seeing is illusory.

From cognitive system to stimulus

We can bring the top-down (concepts-driven) process to the fore by using figures that we can organise in a certain way. The *Gestalttheorie*, first proposed by Christoph von Ehrenfels[5] on melody perception, but elaborated by Max Wertheimer,[6] claimed that the way we perceive parts of a whole depends on the whole configuration of the stimulus. The top-down process, based on knowledge, is clearly seen with the example of the first page of the *Libération* journal published February, 11th, 1986, in which we can read «MAFIA: L'ITALIE JUGE SES PARRAINS» without notice that the first "I" of "Italie" is missing. We have what gelstaltists called a *global* perception of the stimulus, and we not have been influenced by some particular features of it. By way of an example, the picture "Camouflage" is composed by a set of dark blots; at first, we can't tell what this picture depicts, but if we continue or given a hint, fragments are perceptually reorganised and recognised as a Dalmatian dog. The *Gestalttheorie*, or *Form Psychology*,[7] proposed some

*The images descripted in the text, and numerous other ones, are visibles at the specialised web site http://illusionworks.com/html/hall_of_illusions.html.

rules of perception to explain our way of perceiving stimuli as an organisationed construction.

Another kind of researches shows the relevance of background knowledge in the perception processing by using *ambiguous figures*, figures that can be perceived very differently in more than one way. Two of the most famous such figures* was devised by E.G. Boring,[9] «My wife and my mother-in-law»** and Jack Botwinick,[11] «My husband and my father-in-law». We see one or other of the two possible interpretations, and we arrive generally to perceive the other one *when we have the information that another interpretation exists.*

Finally, researches about our automatisms are very instructive too. Indeed, we can't help perceiving a group of letters, or a word, instead of some colour splash on a wall or painting.*** As we are living in a three-dimensional world, we can also perceive in 3D a drawing using perspective ingredients in favour of our perceiver's experience. This probably biologic capacity[4] we are equipped allows psychologists or artists to create a lot of pictures in which what we know about the world and what we see on the drawing enter in conflict, creating an illusory situation. Artists, like Maurits C. Escher,[12,13] usually incorporate two-dimensional depth cues to convey realistic or unrealistic three-dimensional scenes. The most popular such illusion is the "Impossible Triangle", devised by the Swedish artist Oscar Reutersvärd in 1934 and by Roger Penrose:[14] the figure clearly shows that when is given a chance to our visual system to interpret a drawing or image as three-dimensional, it will do so.

Perception is not a reality tracing

All these examples show us that stimuli features and conceptual knowledge can orient our world's perception. The two processing are not separated when we are confronted to stimuli which are not conceived deliberately by researchers. But some such pictures can show the processing interaction: in «Terror Subterra»,[4] two men are in a tunnel; they are strictly identical, but the one who runs after the other is seen large *and* terrific, while the other is seen small *and* terrified.

Perception is then not a real tracing. But his role is major in our attempt to survive, to adapt ourselves to situations, in particular when we are able to perceive resemblances, similarities, analogies between situations or behaviours.

ANALOGY PERCEPTION AND MENTAL FLUIDITY

Analogy perception

Consider the example of the wasp Sphex borrowed to Dean Wooldridge,[15,p.70] (quoted by Hofstadter[16,p.360]):"When the time comes for egg laying, the wasp Sphex builds a burrow for the purpose and seeks out a cricket which she stings in such a way as to paralyse but not kill it. She drags the cricket into the burrow, lays eggs alongside, closes the burrow, then flies away, never to return. In due course, the eggs hatch and the wasp grubs feed off the paralysed cricket, which has not decayed, having been kept in the wasp equivalent of a deepfreeze. To the human mind, such an elaborately organised and seemingly purposeful routine conveys a convincing flavour of logic and thoughtfulness - until more details are

---

*See also the «Duck-Rabbit» figure.[8,4]

**The drawing is inspired by a drawing of W.E. Hill,[10] who has copied an 1888 anonymous German postcard (cf. note on http://illusionworks.com/html/hall_of_illusions.html).

***A good example is the painting of Jasper Johns, «False Start», 1959, which mixes colored splashes and colored words.

examined. For example, the wasp's routine is to bring the paralysed cricket to the burrow, leave it on the threshold, go inside to see that all is well, emerge, and then drag the cricket in. If the cricket is moved a few inches away while the wasp is inside making her preliminary inspection, the wasp, on emerging from the burrow, will bring the cricket back to the threshold, but not inside, and will then repeat the preparatory procedure of entering the burrow to see that everything is all right. If again the cricket is removed a few inches while the wasp is inside, once again she will move the cricket up to the threshold and reenter the burrow for a final check. The wasp never thinks of pulling the cricket straight in. On one occasion this procedure was repeated forty times, always with the same result."

The conclusion of Douglas Hofstadter is that we are not *sphexian*: we can generally realise that we have the same repetitive behaviour. And this is the capacity to perceive resemblances that is the core of conceptual adaptation. It allows us to adapt ourselves to situations which are never identical in a world always changing. Perceiving resemblances is *to perceive a thing as another thing*, consequently perceiving analogies.

An appropriate example is the fact that we have no problem to recognise a letter in spite of its vast number of possible representation. You can represent, for example, some "A" letters feasible with different typefaces, but remind you all the handwritten possibilities. The difficulty is to define exactly the properties of the capital "A".

Hofstadter took an interest in this problem by trying to find the *Spirit* (its alphabet style, e.g. "Times") and the *Letter* (its category, e.g. "A") of a type. Another way to formulate the problem is to conceive a program - LETTER SPIRIT[17,18] - capable to create the alphabet from some letters, i.e. a program capable to "perceive" what is common between some letters and to apply this "spirit" to a new letter.

Conceptual slippages

Analogy perception intervenes each time a conceptual adaptation is necessary. The ability of a concept to be used in unanticipated contexts and manners is possible because we have a gift to make what Hofstadter denotes *conceptual slippages*: such slippages are produced when a "mental representation" slips into a closely related one under sufficient pressure.[16,19,20] Analogy-making is then viewed as the process of recognising that two quite different-seeming situations are actually the same, as long as certain slippages are made depending on mental pressures evoked by the current context.

A good example of conceptual slippage is given by Hofstadter.[21] Consider the "First Lady" concept and answer to the question: Who *in England* corresponds to Nancy Reagan* in the United States? Webster's tells us that it is "the wife or hostess of the chief executive of a country". The implicit pressure represented by a tendency for a same-sex mapping leads for the counterpart of Nancy Reagan to be either Queen Elizabeth or Margaret Thatcher. But Nancy Reagan was not at the centre of the political power, so either Prince Philip or Denis Thatcher (Margaret Thatcher's husband) are really representative of the

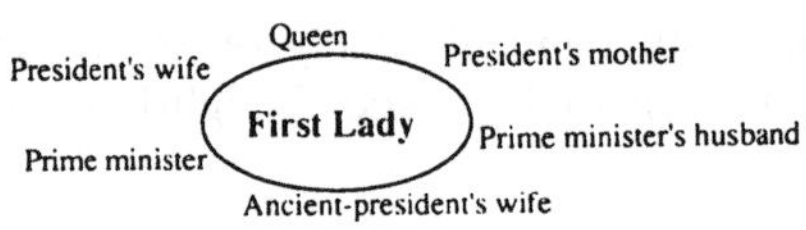

Figure 2. Who in country X corresponds to the First Lady?

*Keep in mind that this was all happening around 1980.

"First Lady" concept! If we continue our journey around the world (Figure 2), no defining rule of the First Lady concept can be elaborated to cover all past, present and future cases.

Depending on the pressures involved, the central-tendency definition can slip very far and yet still seem perfectly natural. It's not what a concept should be that it is important, but what a concept evokes in the people minds. The concept "First Lady", like all concepts, can be adapted to a vast array of situations that no single, rigid definition can over in advance: in other words, concepts are *fluids*.[22,23]

Mental fluidity

People are inclined to make such slippages very naturally in mundane conversation, in particular in the spontaneous production of counterfactual scenarios where things "could have happened" or "ought to have happened". Mental fluidity is then not an exceptional behaviour, and is present in a lot of everyday cognitive activities:* counterfactuals like "If I were you, I would come back to me", or "Suppose I am a hungry hyena and you are a bone!" (in Carroll's *Alice*); rumours-diffusing; word-blending (blend two words to make a new one, like *chocoholic*, or *compuvision* and *infomercials* - two examples given by Levialdi, this volume); analogies like "I am your father for today"; metaphors or figurative sense-using, like "Men are wolves"; pluralisations; recognition (of forms, persons, places, events...); learning (of concepts, savoir-faire, skills...); ideas association, language error-making, slips; all expressions like "Me too", "That's happened to me", "People like me", "Do the same thing as I did"; advertisement-understanding. According to your job or leisure activities, mental fluidity is present in: imaginary scenario (actors, children, role-playing games); text-translating or text-contracting; politics; research and invention; advertisement-making.

Maybe the translation can be considered as the very sophisticated form of analogy-making,[25] if the resolution of an analogy is seen as a translation (or adaptation) from a structure to another.[19] Moreover, when we read a text, the meaning changes between the start and the end of the text. There is no obviousness of sense, so conflicts between critics in literature: it's because there is mental fluidity that there is literary analysis, that is to say *an analysis of the hidden*.

Creativity

Analogy perception can be seen as the capacity to recognise what is common between situations, a sort of core shared between entities or events, in short: *a theme*. According to Hofstadter,[26] creativity is the result of *variations on a theme*. New horizons can then be opened up, either by structures modifications or by working around theme's boundaries. Hofstadter himself reports his experience in the creation of the metaphorical music or *volute* art.[19]

In the field of creativity, mental fluidity plays a major role in: poetry (in particular in the making of analogous wordplay in a different language, or in rewriting a poem without "e"...); art styles-recognising or -creating, or both (e.g., playing same music differently, like a classic piece played as a rock one, or playing a saxophone player Coltrane's piece without saxophone); ambigrams-making (a form of visual play with letters and words);** science-

---

*A lot of examples were given within the context of Lectures presented in particular by Douglas Hofstadter and Gilles Fauconnier at «A Lectures Series in Cognitive Science: Metaphor and Analogy», Povo, Trento (Italy), 18-21 june 1997 (see Fauconnier[24] for examples).

**An example is the mirror-reversible ambigram by Scott Kim, entitled Richard Gregory, which can be read inversed (cf. note on http://illusionworks.com/html/hall_of_illusions.html).

fiction; parody; painting;* humour (jokes, puns, cartoons, caricatures, Monty Python skits...).

Consider for example this humoristic analogy:** Why God won't get tenure? Only published one book. It was in Hebrew. It had no references. He did not publish it in referenced journals. Some doubt He even wrote it Himself. He is not known for His cooperative work. Sure, He created the world, but what has He done lately? He did not get permission from any review board to work with human subjects. When one experiment went awry, He tried to cover it up by drowning all the subjects. When sample subjects do not behave as predicted, He deletes the whole sample. He rarely comes to class, just tells His students to read the Book. It is rumoured that He sometimes lets His Son teach the class. Although He only has 10 requirements, His students often fail His tests. He expelled His first two students for learning. His office hours were infrequent and usually held on a mountain top.

Conclusion

These examples point out the fact that we have a general cognitive capacity of conceptual adaptation or conceptual slippability. To perceive or create analogies is then a common activity. Unexpected conceptual horizons can open up towards ways that are worthy of being explored. The artistic creativity can then be viewed as the ability to perceive the *good* variations. Consequently, the nature of analogy-making is seen as a *general* cognitive process rather than an exceptional mechanism brought to bear only in unusual circumstances. Moreover, mechanisms underlying creativity are mechanisms implied in analogy perception and creation. Since 1978, Hofstadter and his colleagues*** have elaborated several projects to study more rigorously these mechanisms - see Hofstadter[29] for the different arenas observed, and Hofstadter et al.,[22] for a general survey of COPYCAT, JUMBO, LETTER SPIRIT, METACAT, NUMBO, TABLETOP. We are now focus on the COPYCAT one.

THE COPYCAT PROJECT: THE IMPLEMENTED MECHANISMS

The project

COPYCAT[30] - be a copycat means "Do the same thing as I did" - is a computer model of analogy-making implemented by Melanie Mitchell.[31,32] The particularity of this project is to consider that the crux of human analogy-making is abstract (or high-level) perception, i.e. perceiving a thing in a certain way, filtering an abstract essence from a given situation which can then be applied to another situation. The architecture of the model is based on an interaction of a large number of perceptual agents with an associative, overlapping, and context-sensitive network of concepts. This architecture allows such a mechanism of high-level perception that is necessary to give an account of our mental fluidity. The problems it has to resolve have to be resolved in the same way humans do, and the program is capable to give a spectrum of diverse answers to each different problem. Consequently, COPYCAT

---

*Salvador Dali and René Magritte, for example, have often used features belonging to more than one element or object; Solé[27] really paints blended scenes, mixing animals and objects.

**Gilles Fauconnier has presented this humoristic analogy at Povo (see Fauconnier[24] for examples). A good another example is the *doggy* version of Britain's top girl band Spice Girls, *The Spice Grrrs*, for a new ad campaign (cf. *The Sun*, July 8, 1997, with the subtitle "The Sun is No1 for Growl Power"). Humor is a very important indication of the aptitude we have to make conceptual slippage:[28] indeed, we generally understand with a lot of *easiness* humoristic jokes or advertisements.

***At least Daniel Defays, Robert French, James Marshall, Gary McGraw, Marsha Meredith, Melanie Mitchell, John Rehling.

is not about simulating analogy-making *per se*, but is a thoroughgoing exploration of the nature of mental pressures, concepts, and their deep interrelationships, focusing particularly on how pressures can engender slippages of concepts into "neighbouring" concepts - in order to answer with the appropriate manner to the question's problem "Do the same thing as I did", where "same" is the slippery term.

The microworld and its relevance

Inspired by the methodology of physics, Hofstadter has adopted a philosophy of idealising the cognitive phenomena of interest, and crystallising them in a microdomain, where, because of the simplifications, their important properties emerge more clearly than in their full complexity.[29]

The microworld is composed by a no circular alphabet letters, each letter having a knowledge of his neighbour letter. Creative analogy problems are created on this base as «suppose the letter-string **abc** were changed to **abd**, how would you change the letter-string **mrrjjj** in "the same way"?»* This world reduction is necessary to grasp in a clear way all the operations that are used between the perception of a creative analogy problem and the problem's resolution. The analogies are creative in the sense that the problem can have more than one coherent solution, depending on the perception we have of the transformation between the first string and the second string of the problem (Figure 3).

What is happened in the transformation of **abc** in **abd**? The **c** is changing? The *third* letter is changing? The *last* letter is changing? The *higher* letter of the alphabet is changing? And then, what is the element of the string **mrrjjj** which corresponds to the **c**? Is the *last* letter **j**? The *third* letter **r**? The *last group* of letter **jjj**? But what is exactly the transformation: is the **c** (or third letter, or last letter) becomes a **d**, or becomes a *successor* of **c**, or a *successor* of the last or higher letter? All these considerations lead to some solutions, but the given solution depends on the *perception* of the problem - for example, "the last letter becomes **d**", that can lead to the solution **mrrjjd**. One other response is **mrrjjk**, by matching the **abc**'s letter **c** and the **mrrjjj**'s last letter **j**, and then applying the (perceived) rule "replace last letter by successor". Another higher abstract level of perception is to consider neither the letters nor the letter groups (for example, giving **mrrkkk**), but the group length: perceiving **mrrjjj** as the length-string **1**(m)-**2**(r)-**3**(j) leads to lenght term response **1**(m)-**2**(r)-**4**(j), so to letter-string **mrrjjjj** with the rule "replace length of last group by successor".

The COPYCAT model[31] is able to give a solution to a problem depending on the perceived relations and to give another solution to the same problem if the perception of the relations the program "have" is different at the beginning of the resolution.

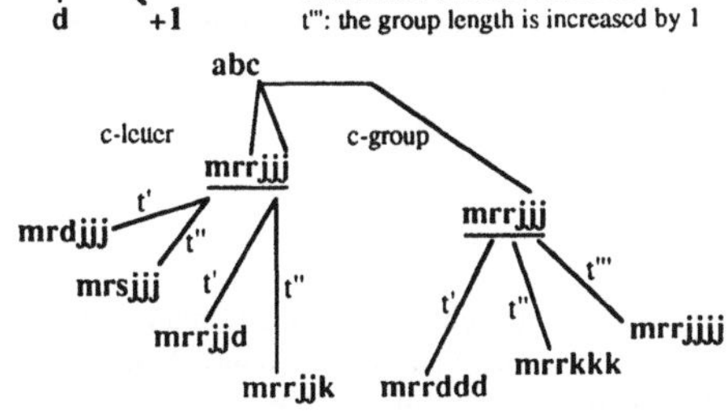

Figure 3. Some solutions to the «**abc** >> **abd**, **mrrjjj** >> **?**» problem, depending on the perceived transformation and the element of **mrrjjj** on which the transformation is applied.

*We note henceforth a problem like this: «**abc** >> **abd**, **mrrjjj** >> **?**».

The model

The COPYCAT cognitive architecture is composed by three major components (see Mitchell,[31] Hofstadter and Mitchell[32] for details). The SLIPNET,* a network of interrelated concepts, is the site of concepts *types*. A numerical length link rely each concept and determines, depending on the "distance" between concepts - distances which can change over the resolution - what slippages are likely and unlikely. The WORKSPACE, the locus of perceptual activity, contains *instances* of various concepts combined into temporary perceptual structures. Finally, the CODERACK contains small agents (the codelets) stochastically selected to carry out tasks in the WORKSPACE - there is several types of codelets depending on different types of structure, and all proposed structure to be build is dependent to a probabilistic decision. Given the **mrrjjj**'s problem, for example, at least the concepts *leftmost, middle, rightmost, letter, letter category* and *string position* acquire a maximal level of activation in the SLIPNET. Codelets begin to explore possible structures as building a bridge between the two last **j** of the string **mrrjjj** or between the **abc**'s **a** and the **mrrjjj**'s **m**. Constructing structures actives other concepts (as *group*) implied in the description of the new situation in the WORKSPACE. Afterwards, the varying activation of concepts orients the selection of the codelets from the CODERACK, so structures to be built.** The activation of the concept *group* can be risen by the perception of another **j** which can be added to the **jj** group. If the three **j** of **mrrjjj** are grouped and linked to the **abc**'s **c**, there is a tendency to perceive **rr** as a group and even **m** as a group, so linked these two groups to respectively **b** and **a** of the **abc** string. Gradually, correspondences are built between **abc** and **abd** and **abc** and **mrrjjj**.
A rule is obtained for the first relation (the **abc-abd** change), *e.g.* "replace letter-category of the rightmost letter by successor", and another rule is adapted to **mrrjjj**, *e.g.* "replace length of rightmost group by successor". The slippages in this case are from *letter-category* to *length* and from *letter* to *group*. The interaction between the perceptual level and the conceptual level, related to an every time probabilistic decision, makes COPYCAT capable to provide different responses to the same problem - for this one, it gives **mrrkkk** (70,5%), **mrrjjk** (20,3%), **mrrjkk** (4,5%), **mrrjjjj** (3,9%), **mrrddd** (0,6%) and **mrrjjd** (0,2%).

Two mechanisms are implemented in the model to give an account of the flexibility of the COPYCAT program. First, a *high-level perception* mechanism is used to give an account of the encoding in a certain way of the problem: perceiving, for example, that it's the third or last letter (rather than the **c**) that is changing in «**abc** >> **abd**», perceiving **mrrjjj** as **m-rr-jjj** (by matching same letters) rather than a series of letters as **m-r-r-j-j-j**, perceiving the string divided as **mrr-jjj** rather than **mr-rj-jj**, and so on. Second, a *perception-conceptualisation loop* is used to give an account of the resolution: for example, perceiving the groups **rr** and **jjj** in the string **mrrjjj** entails to conceive **m** as a group; conceiving **m** as a group of one letter entails to perceive **rr** as a group of two and **jjj** as a group of three; perceiving **jjj** as a group of three letters entails to conceive **4**(j) as a successor of **3**(j), leading to the response **mrrjjjj**.

---

*See the Figure 3.4., Mitchell.[31,p.47]

**This orientation is related to a measure of the coherence of the system in the construction of structures: the *temperature*, varying from 100 (at start) towards 0, lowers as coherent structures are built; this fall orients too the activation of concepts that are adapted to the built structures. It permits to the system to go towards a solution (to be more deterministic); a competition with another orientation can appear and is probabilistically decided; if the second orientation is "prefered", then the temperature rises (the built structure is not coherent with the previous one) to permit another way to solve the problem. Thus, the program gives reasonable and insightful solutions most of the time without being no capable to produce bad or crazy solutions (see Mitchell,[31] Mitchell and Hofstadter[33] for more details).

Conclusion

This particular ant hill-architecture[34] permits the emergence of a robust high-level behaviour from the interactions of a great number of low-level nondeterministic perceptual micro-agents. All the decisions are probabilistic, so at every time the system can move towards a solution or another - depending on all the perceived or constructed features that are more or less leading towards a specific solution, with no *determined* solution. According to Mitchell,[31,p.229] «the program has to have the potential to follow risky (and perhaps farfetched or even crazy) pathways in order for it to have the flexibility to follow subtle and insightful ones». This probabilistic dynamic provide to the model the capacity, by the number of possible solutions given to the same problem, to appear more flexible that the majority of cognitive models. The recent extension METACAT developing by Marshall[35,36] seems to permit the creation of rich representations of the analogies made in this microworld.

COPYCAT is a computer program designed to be able to discover insightful analogies, and to do so in a *psychologically* realistic way.[32] Ideas underlying the project stem from years of close psychological observation in many different arenas, but these observations are not the results of systematic psychological experiments.[29]

Our project, COPSYCAT, is elaborated to show the psychological relevance of the COPYCAT project.

THE COPSYCAT PROJECT: THEIR PSYCHOLOGICAL RELEVANCE

The COPSYCAT project[37] is the examination of the psychological plausibility of the mechanisms postulated in the COPYCAT model to give a psychological account of the conceptual fluidity appearing in numerous activities. The first experiments[38] show that the micro-world material used by subjects is a real material that can exhibit the mental fluidity of subjects on this microworld. The material used to produce creative analogy problems permits more than one solution, so it permits to study which solution is produced by subject and which perception permits to produce it. The experiment presented here shows the reality of high-level perception and perception-conceptualisation loop on this material.

Experimental preamble: microworld and macrosubject

These two mechanisms seem not to be two specific microworld mechanisms that only appear in the COPYCAT's world. The microworld in which COPYCAT discovers analogies is very small but surprisingly subtle. In the First Lady case, our response to the question depends on the way we perceive what is changed between a "chief" and his "first lady", you must just apply the transformation in the new country. But in the creative analogy problems, the same things happen: our response to the problem depends on the way we perceive what is changed between the first and the second string (**abc** and **abd** in the example), you must just apply the transformation to the new string (Figure 4). The new string, as the new country, leads to pressures which orient the perception of the new element which is the object of the transformation.

Another example of the range of this microworld is given by Hofstadter and Gabora.[28] They compare some responses to the «**abc** >> **abd**, **ijk** >> **?**» problem to some responses to a famous joke.

In the problem, the transformation could be designed by the rule "replace the rightmost letter by its successor", yielding to **ijl**. But the change can be took more literally, as "replace the rightmost letter by **d**", yielding to **ijd**. A more literal-minded answer would

Figure 4. abc >> abd, mrrjjj >> ?

be **abd**, based on "replace the entire string by **abd**". The analogies with the following well-known joke is:

- American: Look how free we are in America - nobody prevents us from parading in front of the White House and yelling, "Down with Reagan!"
- Russian: We in Russia, are just as free as you - nobody prevents us from parading in front of the Kremlin and yelling, "Down with Reagan!". **(ijd)**
- Russian: We in Russia, are just as free as you - nobody prevents us from parading in front of the Kremlin and yelling, "Down with Gorbachev!". **(ijl)**
- Russian: We in Russia, are just as free as you - nobody prevents us from parading in front of the White House and yelling, "Down with Reagan!". **(abd)**

Remark that a perfect analogy is not humoristic; it emphasises the conceptual fluidity we can easiness exhibit - it is when the slippage is not made that the joke, *i.e.* the apparition of an event that is not in accordance with our mental habits, exists.

The creation of a microworld has another advantage: the elimination of the ELIZA effect in cognitive models, e.g. english - form symbols mislead us to a computer interpretation of the human world which not exist. In the joke's example, *we* map somewhere "Down with Reagan" with "**d**". COPYCAT is only working on the microworld, although the **d** can be replaced by "Down with Reagan" - but the replacement does not give to the program the *meaning* of the expression[39] (see also Liénard, this volume). The project is really a general one that is not constrained by a domain or a set of hand coded representations - the central point is the *cognitive architecture* of the model which permits an emergent conceptual behaviour taking root in perception[40,41] (see also Adorni, this volume).

The choice for taking first of all this material for real experiments is then motivated by two reasons: i) this material clearly exhibits mechanisms used in mental fluidity with no separation between perception and conceptualisation - the challenge is to increase the explaining power to more semantics material; the subjects that were be given the COPYCAT tasks don't answer in the same way because they can do this with his material. Our outcome is to show that they do this naturally, but that they can apply one simple rule when it is necessary; ii) the subject's behaviour is not reduced - having possibilities to understand the problem in the way he feels it permits to the subject to be free related to his interpretation of the problem. Moreover, this material implies a small base of knowledge (the 26 letters alphabet one) which permits at once to compare seriously subject's resolution and to consider deeply each individual performance.* We finally operate here a world reduction rather than a subject reduction. We sum the actual results of the project - the complete experiment will be a subject of a paper.[42]

*Other criticals have been already formulated.[37]

Experiment

Eighty Université de Provence undergraduates took part in the experiment. For each of them, 10 problems have to be resolved, and for each of the problem, 12 solutions have to be evaluated (Table 1), with computer presentation.

Table 1. The problems and the proposed solutions.

Problems	Solutions "True"	Solutions "False"
lmn >> lmo, kji >> ?	kji kjo kjj kjh lji	ijk kij lkj jkl blo xwf kjk
ijk >> ijl, lmfgop >> ?	lmfgop lmfgol lmfgoq lmfgpq lmfgqr lnfhoq	lmfgqq ijlgoq nohiqr kmfgop lmefop lmfgoz
abc >> abd, abbccd >> ?	abbddd abbcce abbcde abbcef	aababe aaaaad aababx uububc aahahc abbccd babcbd aacacd
aabc >> aabd, ijkk >> ?	djkk jjkk hjkk ikkk ijkd ijkl ijdd ijll	iijl iijd jkkk ijlk
abcd >> abcde, mlkji >> ?	mlkjie mlkjij mlkji mlkjih nmlkji mlkj lkji	abcdi mlkjii fghij mljjk abcd
abcm >> abcn, rijk >> ?	rnnn rijn rijl rjkl nijk sijk	rikl rhij rijh mijk stuv abcs
rst >> rsu, mrrjjj >> ?	mrrjjj mrsjjj mrrjju mrrjjk mrruuu mrrkkk nrsjjk mrrjjjj	mrrklm mrrppp mrrjjf orrjjj
mrs >> mrt, iiii >> ?	mrt iiit iiij iitt iijj jjjj iiiii	ijkl mrss tttt iiim mri
ooe >> o, riippp >> ?	r i p ip ipp	rip rr pp iippp ppp ii rrip
eqe >> qeq, aaabccc >> ?	qeq bbbacbbb baaacccb abbbc	cccbaaa cacbcac abc bbbabbb qqqeppp abcba qaaaecccq bacb

Clearly, the subject is given a problem, gives a solution with no time limit, and evaluates one by one 12 randomly presented solutions for the running problem (maybe his one) with no time limit, clicking on "True" - i.e., "I think it is a possible solution to the problem" -, or on "False" - i.e., "I think it is not a possible solution to the problem". The "True (T)" and "False (F)" propositions varied between 4 and 8 for each problems, so 50% of the two types are presented if we consider all the problems. In sum, 60 T and 60 F propositions are presented to each subject. We suppose that all subjects have the same alphabet knowledge.

The letter-string can be seen at different processing levels (or different levels of abstraction): as a *whole* (W), as a *sequence of letters* (L), as *groups of letters* (G), as *alphabet characterised* (A) and as a *length groups of letters* (LG). Consider the example of problem 7 «**rst** >> **rsu, mrrjjj** >> **?**»:

- if **mrrjjj** is seen at level W (**mrrjjj**), answer could be **mrrjjj** (*no* change - **rst** is different from **mrrjjj**) or **rsu** (*all* string gives **rsu**);
- if **mrrjjj** is seen at level L (**m-r-r-j-j-j**), answer could be **mrujjj** or **mrsjjj** (the *third* letter changes), **mrrjju** or **mrrjjk** (the *last* letter changes);
- if **mrrjjj** is seen at level G (**mr-rj-jj** or **m-rr-jjj**), answer could be **mu-ru-ju** or **ms-rk-jk** (the *last letter of each group* changes), **m-rr-uuu** or **m-rr-kkk** (the *last group of letters* changes) or **n-rs-jjk** (*each last letter of each group* changes);
- if **mrrjjj** is seen at level A (the alphabetic features are considered, as the alphabetic jump between letters **m** and **r** and **r** and **j**, the alphabetic direction from **m** down to **r** and from **r** up to **j**, and so on), answer could be **mrryyy** (same *gap* between considered letters, *i.e.* **m**-n-o-p-q-**rr**-s-t-u-v-**www**).
- if **mrrjjj** is seen at level LG (**1m-2r-3j**), answer could be **mrrjjjj** (**1m-2r-4j**).

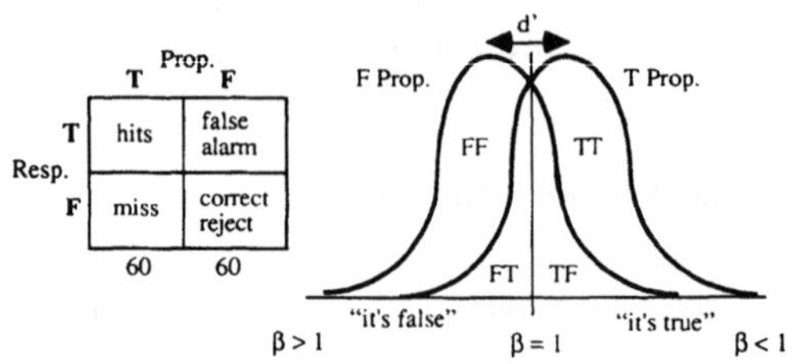

Figure 5. The adapted SDT stimulus-response matrix with the two indices d' and β.

Three factors are manipulated during the experiment: 50% of subjects have three examples before the 10 problems, 50% of the subjects have the problem reminded before each 12 evaluations, and 50% of the subjects have the problems in a randomly presentation - so, 10 people in each of the eight formed conditions. As these factors seem not to intervene in our attempt to demonstrate the two mechanisms previously evoked, we not go back over in this chapter. Remind just that problems are *not always* under an ordered presentation (showed in the Table 1) and that all propositions that have to be evaluated are *always* under a randomly presentation.

At the beginning of the experiment, a subject gives his response to the problem. Then, 12 propositions are presented one at a time. He can answer T or F to a T or F proposition. Consequently, we have four possible cases that are easily analysed by the *Signal Detection Theory*[43-45] or SDT (Figure 5). This frame of analyses permits to use two indices: the *discrimination index* (d') and the *decision index* (β).

According to this model, a subject's ability to discriminate between true items and false items is given by d', the distance between the means of the true and false distributions in units of the common standard deviation. The β criterion measures the subject's criterion of decision, that it is: does he prefer raise the risk to miss hits (i.e., to respond F to a T proposition) or to be directed towards false alarm (i.e., to respond T to a F proposition)? The case in which d = 0 and β = 1 corresponds to chance decision. These two indices, on which means can be calculated without neglect the global variations, are obtained by measuring the proportions of hits (i.e., TT) on the total of T (60) and the proportions of false alarm (i.e., TF) on the total of F (60).

Hypotheses and results on the high-level perception mechanism

The hypothesis is that subjects don't perceive all the possible solutions for a problem, that is they don't perceive all the relations that permit to produce all these responses.

The diagram (Figure 6) shows the position of all subjects (**S**) within the space on which there is the *machine-behaviour* (**M**) - a "no-error behaviour" - and the *chance-behaviour* (**C**) tables. **M** represents a subject who makes no error (d' —> ∞; β = 1). **C** represents a subject who responds with no criterion of decision (d' = 0, β = 1). The two βs are the same because the two distributions in each case are symmetric: in the case of **M**, the distributions are very distant and in the case of **C** they are astounded. **S** represents the set of subjects (d' = 2,164 [s = 1,518], β = 0,522 [s = 0,371]). We show a table of a subject as an illustration of the set (d' = 2,247, β = 0,532). These global results show that there is a selection of True propositions *that is not a chance selection.* The d' index is high enough to

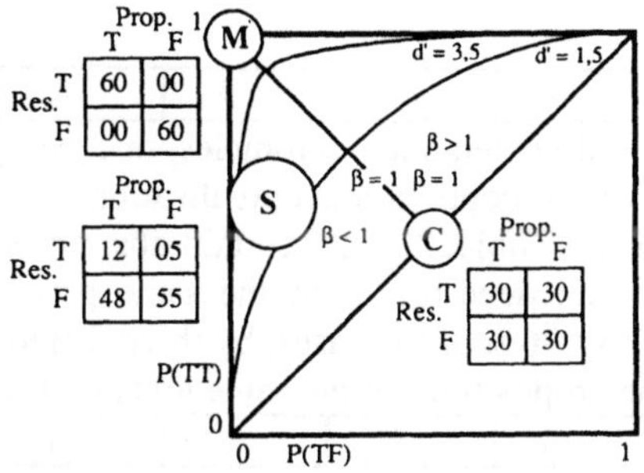

Figure 6. Global results in the form of a ROC (relative operating characteristic) with examples of a chance-behaviour (C), of a subject-behaviour as an illustration of the set of subjects (S) and of the no-error machine-behaviour (M).

say that the two distributions T and F are well differentiated by subjects. The $\beta < 1$ shows that subjects have rather judged the "true-lity" of the propositions.

Then, we can conclude of the existence of the high-level perception on his material, as it was defined before: *the string is perceived at a certain abstract level, even if perception time is not limited.* What happens after this oriented perception?

Hypotheses and results on the perception-conceptualisation loop

In principle, the loop implies that responses or evaluations made by subjects can influence other responses and evaluations they have to do. We have organised the results in the form of chronicles. For example, a subject has to solve the problem 5 - «**abcd** >> **abcde, mlkji** >> **?**» - in tenth position (the problems can be presented in random order) and has to evaluate the 12 corresponding propositions (cf. Table 1). The propositions are randomly presented as follow for this subject (Table 2).

Table 2. Example of a problem resolution with the presentation order of the propositions. The bold letters represent the true propositions perceived by the subject. The propositions judged false are not taken in account here.

	propositions (n° in Table 1)	true/false (subject's response)	levels (true propositions)
subject's answer	mlkjij	(T)	**L**
1st proposition	mlkjij (2)	T (T)	**L**
2nd proposition	mlkjih (4)	T (T)	**A**
3rd proposition	abcd (12)	F (F)	-
4th proposition	lkji (7)	T (F)	A
5th proposition	mlkji (3)	T (F)	W
6th proposition	mlkj (6)	T (F)	A
7th proposition	mljjk (11)	F (F)	-
8th proposition	nmlkji (5)	T (T)	**A**
9th proposition	mlkjii (9)	F (T)	-
10th proposition	fghij (10)	F (T)	-
11th proposition	abcdi (8)	F (F)	-
12th proposition	mlkjie (1)	T (F)	L

The chronicle is:

L	**L**	**A**	A	W	A	**A**	L

Each subject is characterised by ten such chronicles, one per problem. Then, we can analyse some implications of the perception-conceptualisation loop.

The hypothesis is that when a subject is at a certain abstract level X, he perceives the other propositions which can be judged true with the same processing level X. Within a problem, we obtain this in almost 70% of the cases, by three different ways:

i) Subject perceives all the propositions at the same level (9,4%):

L	**L**	G	G	**L**	LG	G	W

ii) Subject perceives some propositions at the same level (43,9%):

A	L	**A**	A	**A**	W	L	A

iii) Subject perceives all propositions at the same level and others (16,4%):

G	LG	W	**L**	**L**	**G**	L	**G**	**G**

If we consider two problems at time and analyse the impact of an abstract level (at the first problem) on the responses and evaluations (at the second problem), subjects are in 74% of the cases at:

i) the same level:

L	L	**L**	G	G	G	G
L	G	G	G	G	**L**	

ii) the lower level - the first level can not exist in the second considered problem:

G	L	LG	**G**	**G**	L	**G**	W	L
L	A	T	**L**	L	A			

These last results have to be analysed more deeply to verify a prediction made in the COPYCAT project: lower abstract levels are more easily perceived than higher ones, but higher ones, when perceived, disappear more slowly.

Anyway, the perception-conceptualisation loop exists (see also Cardaci et al., this volume): indeed, when one is at a certain abstract level of perception and gives his response, he perceives generally the true propositions which are at the same level. To go further, we have to determine exactly what is the implication of one response on the following one. We have per subject the matrix of inter-level implication, i.e. the percentage to perceive a Y-level proposition after having perceived a X-level one. This work is in progress - we will use the Bayesian Implicative Analysis[46,47] for the data treatment.

CONCLUSION

Mechanisms as high-level perception and perception-conceptualisation loop are analogy-making mechanisms. But analogy-making is at centre of conceptual fluidity, and conceptual fluidity is characterised by its "ubiquitous" property. So, mechanisms underlying analogy-making are *general* mechanisms. Two consequences follow.

What is a concept? Concepts have to be integrated in mechanisms which permit conceptual fluidity. The *fluid* concepts[22] have to be defined in order to be concepts that permit analogy-making - i.e. that have a propensity to *slip*. Consider "dog":[23] it is generally defined by some properties like "four legs", "mammal", "barking", and so on. But it would be possible that a dog is like a caretaker - it watches the housing. So, "dog" as concept means nothing if we take in account the mental fluidity: "dog" and "caretaker" are two instances of the same concept when we make an analogy between them. If concepts are *rigids*, how can we make so much analogies? A concept appears when different entities are perceived as the same one - it's in fact the application of the proverb "plus ça change, plus c'est la même chose". The final outcome of the COPSYCAT project is to propose a theory of concepts nature and organisation that allows the conceptual slippages arguments.[48]

What is cognitive modelisation? Perception can not be separated of cognition (as defended by Fodor[49]). Generally, perception is the base of the processing, but the *cognitive* processing is what is realised *after* the perception phase. The perception-conceptualisation loop is in fact the corollary of the high-level perception. Indeed, we perceive situations or entities at a certain abstract level, in virtue of what happens before. Concepts can orient perception and perception can orient conceptualisation, but the interaction is not an interaction inter-modules.* According to Goldstone & Barsalou,[51,p.254] «it may be useful to describe a *continuum* from perceptual to conceptual. What varies along this continuum is how much and what sort of top-down processing has been done to bottom-up input information. Specifying exactly where expectations and conceptual pressures influence processing along the perception/conception continuum is a real, although highly empirical, question». The "fodorian approach"[41] permits - or even encourages - to propose models operating on *frozen* representations architecture:[32] we study the cognitive processing *after* the elaboration of the representation.

The cognitive modelisation must be constrained by the perception-conceptualisation loop - *the question is not how many "concepts" are activated, but why these ones are*. What is more, an architecture which not separate perception from conception could permit to conceive general models including a wide range of different possible behaviours in the frame of what is modelisated by the model - in other words, a model which include in addition to "normal" behaviour the "creative" and the "pathological" ones (see Torre,[52] for the same observations in favour of the chaotic approach). This particularity of models capable to modelise fluidity is summed by the property of unpredictability of the issues of the system in front of the problem it has to resolve.

The adapted solution to a problem - a possible definition of creativity - requires to consider all contextual features which guide resolution.[53,54] This human property of being able to change our "conceptual perception" during a resolution is probably reduced in psychological experiments, if subjects have not numerous choice to solve a problem, e.g. are not free to answer "maybe" when are only given "yes" or "no" possible answers. To perceive in a certain way and to be influenced by the *spiral* between percepts and concepts entails a conceptual adaptability in front of very different situations or problems. Such a spiral must be studied to grasp the core of cognition, but is as endlessly as the Roberto Juarroz' definition of poetry:[55] *Vision which create what it sees.*

ACKNOWLEDGEMENTS

I would thank Jean-Sylvain Liénard and Bertrand Zavidovique; François Antoni, Claude Bastien, Abdessadek El Ahmadi, Gilles Fauconnier and Douglas Hofstadter for helpful experimental and epistemological discussions; Philippe Minh, Aline Pélissier, Alain Tête and in particular Jean Cristofol for his welcome at the Villa de l'Ecole d'Art d'Aix-en-Provence where I exposed in march 1998 some of the ideas presented here in a lecture entitled «La créativité? Percevoir des analogies»; Marie-Laure Barbier and Martial Polge for permit to me in may 1998 to expose and discuss the role of analogy in creativity at «Créativité et cognition» (6ème Journée d'Etude en Sciences Cognitives de l'Association ACCION) in a lecture entitled «L'Analogie comme source de la Créativité»; Todd Lubart for his generous participation at this 6ème Journée.

*See in particular Forbus et al.[50] for the "modular" interaction and Chalmers et al.[40] for the "no separated" one.

REFERENCES

1. D.W. Massaro, Visual perception, in: *Experimental Psychology. An Information Processing Approach*, Harcourt Brace Jovanovich Publishers (1989).
2. R.L Gregory, Visual illusions, *Scientific American* 219:5 (1968).
3. R.N. Shepard, Psychophysical complementarity, in: *Perceptual Organization*, M. Kubovy, and J. Pomerantz eds., Lawrence Erlbaum Associates, Hillsdale NJ (1981).
4. R.N Shepard, *Mind Sights: Original Visual Illusions, Ambiguities, and other Anomalies*, W.H. Freeman and Company, New York (1990).
5. C. von Ehrenfels, Über Gestaltqualitäten, *Vtljsch. Wiss. Philos*, 14 (1890).
6. M. Wertheimer, Experimentelle Studien über das Sehen von Bewegungen, *Zeitschrift für Psychologie* 61 (1912).
7. P. Guillaume, *La Psychologie de la Forme*, Flammarion, Paris (1937).
8. J. Jastrow, *Fact and Fable in Psychology*, Houghton-Mifflin, New York (1900).
9. E.G. Boring, A new ambiguous figure, *American Journal of Psychology* 42 (1930).
10. W.E. Hill, My wife and my mother-in-law, *Puck*, November 16:11 (1915).
11. J. Botwinick, Husband and father-in-law: A reversible figure, *American Journal of Psychology* 74 (1961).
12. M.C. Escher, *Grafiek en Tekeningen*, tr. fr., 1993, *L'œuvre graphique*, Taschen, Köln (1959).
13. B. Ernst, *Avonturen met Onmogelijke Figuren*, tr. fr. by M. Gendron, 1990, *L'Aventure des Figures Impossibles*, Taschen, Köln (1985).
14. L. Penrose and R. Penrose, Impossible objects: A special type of visual illusion, *British Journal of Psychology* 49 (1958).
15. D. Wooldridge, *Mechanical Man-The Physical Basis of Intelligent Life*, McGraw-Hill, New York (1968).
16. D.R. Hofstadter, *Gödel, Escher, Bach: an Eternal Golden Braid*, Basic Books, New York (1979).
17. D.R. Hofstadter and G. McGraw, 1995, Letter Spirit: Esthetic perception and creative play in the rich microcosm of the roman alphabet, in: *Fluid Concepts and Creative Analogies*, D.R. Hofstadter, and the Fluid Analogies Research Group eds., The Penguin Press, Somerset (1997).
18. J. Rehling and D. Hofstadter, Letter Spirit: Modeling creativity in a visual domain, in: *Proceedings of the Nineteenth Annual Conference of the Cognitive Science Society*, M.G. Shafto and P. Langley eds., Lawrence Erlbaum Associates, Mahwah NJ (1997).
19. D.R. Hofstadter, *Metamagical Themas: Questing for the Essence of Mind and Pattern*, Basic Books, New York (1985).
20. R.M. French, *The Subtlety of Sameness: A Theory and Computer Model of Analogy-Making*, The MIT Press/A Bradford Book, Cambridge MA (1995).
21. D.R. Hofstadter, How might analogy, the core of human thinking, be understood by computers?, in: *Metamagical Themas: Questing for the Essence of Mind and Pattern*, Basic Books, New York (1981).
22. D.R. Hofstadter and The Fluid Analogy Research Group eds., 1995, *Fluid Concepts and Creative Analogies*, The Penguin Press, Somerset (1997).
23. B. Vivicorsi, "Dog" is not a concept, oral presentation at the *Fifth European Congress of Psychology*, Dublin, July 6th-11th, Ireland (1997).
24. G. Fauconnier, *Mappings in Thought and Langage*, Cambridge University Press, New York (1997).
25. D.R. Hofstadter, *Le Ton beau de Marot: In Praise of the Music of Language*, Basic Books, New York (1997).
26. D.R. Hofstadter, Variations on a theme as the essence of imagination, in: *Metamagical Themas: Questing for the Essence of Mind and Pattern*, Basic Books, New York (1982).
27. J. Solé, *Animaleries et compagnie*, Vents d'Ouest, Rennes (1991).
28. D.R. Hofstadter and L. Gabora, Synopsis of the Workshop on Humor and Cognition, *Humor-International Journal of Humor Research* 2-4 (1989).
29. D.R. Hofstadter, Past, Present, and Future of the Center for Research on Concepts and Cognition at Indiana University, Bloomington (1992).
30. D.R. Hofstadter, The Copycat Project: An Experiment in Nondeterminism and Creative Analogies, MIT Artificial Intelligence Laboratory Memo No.755, January 1984.
31. M. Mitchell, *Analogy-Making as Perception: A Computer Model*, The MIT Press/A Bradford Book, Cambridge MA (1993).
32. D.R. Hofstadter and M. Mitchell, The Copycat project: A model of mental fluidity and analogy-making, in: *Advances in Connectionist and Neural Computation Theory, Vol.2: Analogical Connections*, K.J. Holyoak and J.A. Barnden eds., Ablex Publishing Corporation NJ (1994).
33. M. Mitchell and D.R. Hofstadter, The role of computational temperature in a computer model of concepts and analogy-making, in: *Proceedings of the Eleventh Annual Conference of the Cognitive Science Society*, Lawrence Erlbaum Associates, Hillsdale NJ (1989).
34. B. Vivicorsi, Les glissements conceptuels, ou l'«architecture-fourmilière» comme architecture cognitive, in: *Actes du 6ᵉ Colloque de l'ARC, «Sciences Cognitives, Individus et Sociétés»*, F. Anceaux and J.-M. Coquery eds., Villeneuve d'Ascq, 10-12 décembre, France (1996).

35. J.B. Marshall, From Copycat to Metacat: Developing a self-watching framework for analogy-making, Paper presented at the *Mind II: Computational Models of Creative Cognition*, Dublin, Ireland, September 1997.
36. J.B. Marshall and D.R. Hofstadter, Making sense of analogies in Metacat, in: *Advances in Analogy Research: Integration of Theory and Data from the Cognitive, Computational, and Neural Sciences*, K. Holyoak, D. Gentner, and B. Kokinov eds., NBU Series in Cognitive Science, New Bulgarian University, Sofia (1998).
37. B. Vivicorsi, The Copsycat project: Towards a conceptual fluidity theory, in: *Advances in Analogy Research: Integration of Theory and Data from the Cognitive, Computational, and Neural Sciences*, K. Holyoak, D. Gentner, and B. Kokinov eds., NBU Series in Cognitive Science, New Bulgarian University, Sofia (1998).
38. B. Vivicorsi, Analogies créatives et architecture cognitive: approche expérimentale du modèle Copycat, Post-Master Degree in Cognitive Psychology, Université de Provence, France (1996).
39. D.R. Hofstadter, The ineradicable Eliza effect and its dangers, (1995), in: *Fluid Concepts and Creative Analogies*, D.R. Hofstadter, and the Fluid Analogies Research Group eds., The Penguin Press, Somerset (1997).
40. D.J. Chalmers, R.M. French, and D.R. Hofstadter, High-level perception, representation, and analogy: A critique of artificial intelligence methodology, *J. Expt. Theor. Artif. Intell.* 4 (1992).
41. B. Vivicorsi, De Fodor à Hofstadter, ou d'un mystère à la nécessité de son dévoilement, *Bulletin de Psychologie*, to appear.
42. B. Vivicorsi and C. Bastien, Mechanisms underlying mental fluidity: Experimental evidence in the Copycat microworld, in preparation.
43. J.A. Swets, Indices of discrimination or diagnostic accuracy: Their ROCs and implied models, *Psychological Bulletin* 99 (1) (1986).
44. J.A. Swets, Form of empirical ROCs in discrimination and diagnostic tasks: Implications for theory and measurement of performance, *Psychological Bulletin* 99 (2) (1986).
45. J.A. Swets, W.P. Tanner, and T.G. Birdsall, Decision processes in perception, *Psychological Review* 68 (5) (1961).
46. J.-M. Bernard and C. Charron, L'Analyse Implicative Bayésienne, une méthode pour l'étude des dépendances orientées. I: données binaires, *Math. Inf. Sci. hum.* 134 (1996).
47. J.-M. Bernard and C. Charron, L'Analyse Implicative Bayésienne, une méthode pour l'étude des dépendances orientées. II: modèle logique sur un tableau de contingence, *Math. Inf. Sci. hum.* 135 (1996).
48. B. Vivicorsi, Glissements conceptuels et fluidité mentale: de Copycat à Copsycat, Doctoral Dissertation, Université de Provence, France, in preparation.
49. J.A. Fodor, *The Modularity of Mind. An essay on faculty psychology*, MIT Press, Cambridge, MA (1983).
50. K.D. Forbus, D. Gentner, A.B. Markman, and R.W. Ferguson, Analogy just looks like high level perception: Why a domain-general approach to analogical mapping is right, *J. Expt. Theor. Artif. Intell.*, in press.
51. R.L. Goldstone and L.W. Barsalou, Reuniting perception and cognition, in: *Rules and similarity in human thinking*, S. Sloman and L. Rips eds., special issue of *Cognition* 65 (1998).
52. C.A. Torre, Chaos, creativity, and innovation: Toward a dynamical model of problem solving, in: *Chaos Theory in Psychology and the Life Sciences*, R. Robertson and A. Combs eds., Lawrence Erlbaum Associates, Mahwah, NJ (1995).
53. C. Bastien, Does context modulate or underlie human knowledge?, in: *Cognition and Context*, A.C. Quelhas and F. Pereira eds., special Issue of *Analise Psicologica* (1998).
54. B. Kokinov, Analogy is like cognition: Dynamic, emergent, and context-sensitive, in: *Advances in Analogy Research: Integration of Theory and Data from the Cognitive, Computational, and Neural Sciences*, K. Holyoak, D. Gentner, and B. Kokinov eds., NBU Series in Cognitive Science, New Bulgarian University, Sofia (1998).
55. R. Juarroz, La vision qui crée ce qu'elle voit, in: *Sciences et Imaginaire*, I.A. Maréchal ed., Albin Michel, Paris (1994).

PANEL SUMMARY: **CREATIVITY IN 22 QUESTIONS: WHAT'S THE CATCH?**

Bertrand Zavidovique

Paris 11 University
IEF 91405 ORSAY, France
e-mail: zavido@ief.u-psud.fr

INTRODUCTION

This panel session on creativity was meant to address some primal question like *whether Creativity exists as a phenomenon* or *what basic properties could be assigned to it that would be agreed on in both the living and technology realm*, apart from scattered samples of its supposed productions. Such matter is often close to faith or metaphysics in the beginning, thus it is fruitful to consider starting examples from Art, that is from an intermediary and intuitive source between life and technique.

For instance one can be interested in the **Bull's head** by Picasso (see Figure 1a), as a genuine counter example against too quick and trivial answers. The bronze cast is universally considered a piece of pure creativity: yet nothing is truly created, no tools, not even really an object, nothing but the idea of the beast put to a concrete form in the deceptive use of bicycle parts. As a consequence *Mind* appears a potential major actor in creativity, while actual creations might be questioned again as valid proof of its existence. Bulls always fascinated men. Some 15,000 years ago, an artist was already painting the head of a **black bull** (see Figure 1b) on the walls of a cave at Lascaux in Dordogne (France). Was he or she more or less inventive than Picasso? The significant gap in time, culture and state of the technique, together with the remarkable capture of the dynamics of animals by the painter, suggest further thinking matters: time and its contribution, among which are civilization or technology advances, may turn out to have no impact on creativity. This does not make creativity a static process: the question of the time part and dynamics in it is so raised.

Let us consider a last example to make our start point clear: during the seventies in the US a trend of monumental sculptures did widespread, somewhat related to tendencies towards gigantism in other fields like architecture. The **clothes-peg** by Claes Oldenburg in 1967 (see Figure 1c) is officially a mixture inspired from both the Empire State Building

Human and Machine Perception 2: Emergence, Attention, and Creativity,
edited by Cantoni *et al.* Kluwer Academic / Plenum Publishers, New York, 1999.

a) b) c)

Figure 1. a) Unusual bull's head by Picasso. b) Much older and traditional bull's head. c) Clothes-peg in the Oldenburg's style.

and the humble small hanger. According to the author it was so much close to some proposals at the famous Chicago Tribune contest in 1922 that he named the drawing by "late presentation to the architectural contest..." before making a 14 meter high statue in steel out of it. One could see as well something of the **Eiffel Tower** in this erected peg with its two legs, and further be reminded of any metallic similar hanger structure like a crane. Now, here again modern gigantic constructions are often related to very ancient buildings such as the famous 4,000 years old Stonehenge, not to forget that the crane is about 5,000 years old at least and yet a symbol of modern technology. Like Picasso, Oldenburg in some minor sense reaches creativity by diverting the purpose of a known object into another thing, almost negating the very concept of model. There is like a trick which leads to wonder whether creativity could be taught or transmitted. Remains that it seems hard to tell somebody how to invent the clothes-peg though, that the gimmick does not look like inspired from any natural thing or animal, and that a model was likely never designed prior to make it.

From the examples above let us risk non creativity in trying to model Creativity as a set-function or a set-relation χ from A to B, since it produces something somewhere. It was conjectured earlier that set A is meant to be "man", i.e. the set of minds and we assume faithfully that it is not empty. At least four categories of questions pop out as far as the *existence and properties of* χ*, the Creativity,* are concerned:

- does B exist, meaning is it not empty? Being made of creations, is B different from A, i.e. are creations more than theoretical views? Can a map product of such functions be computed, is the composition associative..?
- is $A \subseteq B$ or $B \subseteq A$? Meaning is the function χ internal, then could it be inverted on any part of (A,B)?
- in case recursivity is further proved, does a fixed point exist and can bounds or extremal points be exhibited?
- what does the function χ actually consist of and how can it be represented?

The first category is concerned with properties of the output set, that we call *nature* later for simplification (actually B is rather a set of functions from nature (or mind?) into nature (or mind?) but the distinction does not bring much at that point). The second deals with the input set, tentatively called *mind* already, and with the invertibility of the function χ. The third one addresses the question of set bounds and of optimisation, that sends back to actual *creation*. Eventually the function graph and its representation, or the concept of model, are referred to as *abstraction* in the sequel. These names allow for more concrete images in minds: every category is progressively exemplified by a series of rather mundane questions in which the last one is an opening and a transition towards the next group (part I). The 22nd and last question introduces the paradox to emerge with creativity being a process. The latter "catch 22" situation is explained (part II) under the form of a ritornelle (like these little Christmas songs). That leads to conclude (part III) by a proof: if it belongs to logic, χ, **the Creativity can exist as a dynamic process that allows mind to enhance itself.**

A GAME BETWEEN NATURE OR MIND AND CREATION OR ABSTRACTION (part I)

Nature

1. If nature created Man, Man created students, students create everything ... does nature create everything?

2. Is nature a reference, an absorbing point, a fixed point ... in creativity?

3. Which ones are creative: making a baby, raising a child, cell mythosis, Langdon's automata (re)production ..?

4. Is Life, or Be living, the primal creativity? The ultimate creativity?

Mind

5. If Teaching is a projection onto the learner ('s skills) is Creating any (pseudo) inverse?

6. Does perception precede, influence ... creation?

7. Does creativity need reference, or is it the ability to learn with no reference?

8. Is one of these creative: declination, analogy, generalisation, coding, rephrasing or wording..?

9. Can one, or a system, learn to be creative?

Creation

10. If creation may locally optimise life span and ease of a specie, what are the reward, the incentive, the optimisation ... of creativity?

11. Mankind creates things that change man, do things create man (see Figure 2) in any way?

12. Is creating a mess creative?

13. Is the ability of creating tools a prerequisite in creativity?*

14. Does a machine tool create an object, a compiler a program, a compiler an executable?

15. Are technology, automation ... helping or contradicting creativity?**

16. Technology helps productivity, productivity helps to create more, wouldn't chain-creation be a negation of creativity?

17. Does Creation demonstrate Creativity?

---

*If yes wouldn't it be a (metacircular) definition of it?

**Did computers help making more stories or more books?

Figure 2. Do things create man?

Abstraction

18. If creativity involves abstracting variables, is a population creative in voting?
19. Is creativity the ability to create a model, to create a model first..?
20. Does it take creativity to solve every problem?
21. Are these intellectual processes creative: asking a question, making an error, asking which question was not asked that ought to be?*
22. Does creativity require consciousness of creating?**

CREATIVITY, MIND AND NATURE: A CATCH 22 SITUATION (part II)

Trying to skip any theological or metaphysics debate let «nature» be our universe of discourse:

nature includes everything.

If nature includes it all then

either only part of nature is creative or *nature is the only creative.*

If only a strict subset of nature is creative (let's call the part «living» or «mind», and «perception» is a map from nature onto mind) then

either it learnt from nature or creativity is not learnable,

If creativity is not learnable then it is natural .....and *nature is the only creative.*

If mind learnt creativity from nature then

either there is more into it than mere inversion of perception (let this more be «abstraction» or «understanding») or mind only mimics nature

*Interesting enough: if it would raise the question «when is lunch?» it might appear as having made an error since the panel is not over, but it would still help create lunch in some way.

**An Aristotelian paraphrase: «We do not know, if we do not know that we know.» Can any one or thing create, not being aware of it?

If mind mimics nature then nothing much is created that weren't already here

.....and *nature is the only creative.*

If abstraction yields more than reversed perception then

either mind does not belong all to nature, and this is contradicting our axiom and theological as well,

or mind productions are not all included in nature since:

perception(nature)$\subseteq$mind and perception^{-1}(mind)-nature=abstraction/abstraction$\neq\varnothing$

and then

nature does not include it all.

Neither does perception^{-1}(mind), because:

if it would, then either only part of it is creative or...et caetera, et caetera!

CONCLUSION (part III)

Where not all Questions are Answered. Still, Machines have a long way to go!

Eventually, not to be left with nothing, one could

-state some supplementary dimension starting with time of which a metaphor in this model is iterations of **P** o **P**$^{-1}$ (**P** is perception for short) in an attempt to bridge nature and abstraction,

-state that not all nature is readily perceivable and that some of it may never be, some part of mind to start with and to fit uncertainty principles,

-call faithfully "consciousness" the set-difference between **P**(**P**$^{-1}$(mind)) and **P**(nature),

and then for the ultimate satisfaction of the mind

mind can be the only creative

(as long as it does not include everything!)

creativity can be purely abstract

like the ability of creating models (mind $\cap$**P**$^{-1}$(**P**(nature)) for instance)

mind creates itself (extends, enlarges, enhances...)

and this is the reward of creativity.

Applying **P** and **P**$^{-1}$

mind aims at comprehending nature, but that is unachievable goal according to our model...

because in succeeding it would include everything and, being the only creative, not include it all, et caetera ...

RECOGNISING, REMEMBERING AND KNOWING

Ornella Andreani-Dentici

Istituto di Psicologia
Università di Pavia
I-27100 Pavia, Italy

INTRODUCTION

Memory supports an extraordinary range of functions: remembering pictures, words, autobiographical events, places, learning concepts, learning skills as driving, playing piano, solving mathematical problems and also developing emotional responses such as fear of dogs or strangers.

Trying to understand this system, psychologists and neuropsychologists have developed individual models for each situation supported in this trial by neuroscientists who emphasise cases of dissociation between functions: so we have assisted to an abundance of taxonomies which include Short Term Memory (STM) / Long Term Memory (LTM), visual or verbal memory, explicit and implicit, episodic/semantic, procedural/declarative memory. On the other end, we have *global approaches*, which try to design a general cognitive architecture for a global system that could account for all these functions.

First, it is necessary to limit terminological confusion and excess: Gardiner and Java[1] invite to avoid to use the same terms to label *memory tasks*, *memory systems* or *processes*, *states of awareness*.

They affirm that this practice is not merely confusing, but misguiding because it suggests an identity between the tasks and the systems or processes. They propose the cross classification presented in Table 1, arguing that implicit memory cannot be used as a name of a memory system or a state of awareness, and that conscious remembering cannot be used as a name of memory process.

Table 1. Classification of terms (from Gardiner & Java, Recognizing and remembering, in: Collins and others eds., Theories of memory, 1993).

Category	*Exemplars*
Memory tasks	Explicit, implicit
Hypothetical constructs	System: episodic, semantic, procedural, perceptual representation Process: conceptually driven, data-driven
States of awareness	Conscious recollection or remembering Feelings of familiarity or knowing Unaware: unremembered and unknown

In *explicit* tasks the instructions ask the conscious recollection of some prior study list, in *implicit* memory precedent knowledge is not required, but it does facilitate the task (priming, skill learning, reading speed).

For instance, in the *stem completion tasks* (where subjects are given the initial letters of a word and are asked to complete terms with the first word that comes to mind), subjects more frequently perform the completion using a word presented in a prior list: this facilitation, which is a sign of implicit memory, appears also in severe amnesics.

Another dissociation found in amnesics concerns priming effects in reading: they read faster a text in the second reading, but they cannot answer to questions on the text: so they show perceptual but not semantic priming.

In parallel with the increasing number of experimental paradigms and clinical observations on dissociations, we have assisted to the increase of hypothesis of various memory systems such as STM/LTM, episodic/semantic, procedural/declarative, or to the description of different processes based on unitary systems, such as retrieving, recognising, knowing, remembering, automatical and intentional processing. In the recent years the focus of attention has been centred on the *remember and know* experiments, that is on the variables which affect the states of conscious recollection and the sense of familiarity. In this type of experiments the task requires the conscious recall of a prior experienced event, but the subjects must report if they remember that an item has previously been presented, and can recall it, or if they have only a vague sense of familiarity, without conscious recollection. Differences between normal and amnesics, young and older subjects, velocity of presentation, divided attention, confidence ratings have been examined, showing that the problem does not concerns only the task (explicit or implicit) and the memory system, but it involves the subjective experience, the motivation of the subject, the degree of elaboration, the self-confidence and so on.

But the question is: do the judgements of familiarity result from a partial retrieval of the target or they must be attributed to a pre-storage semantic information? May be that feeling of familiarity is activated by the perceptual system, which is immediate but not conscious?

MODELS OF MEMORY SYSTEMS

Perhaps it is useful to make a brief review of the research on the models of memory systems in their evolution from increasing sophistication in experimental design and from the comparison with data on pathological deficit and effects of brain lesions in experimental animals. In fact, if we want to understand the organisation of cognition we must considerate the functions (psychological level of mental operations), the structure and the mechanisms (neurological and biological level) and if possible introduce all these components in simulation models.

The study of cognitive processes includes as a principal problem understanding how memory processes contribute to the conscious experience of remembering.

Since 1985 Tulving[2] has proposed the existence of different systems for episodic, semantic and procedural memory: conscious recollection or remembering is a by-product of retrieval from the episodic system that is influenced by conceptual processing; feelings of familiarity or knowing are typical of retrieval from semantic system; in the absence of conscious recollection lack of awareness is typical of the procedural and of the perceptual representation system, which show the persistence of learned habits or experiences without subject's active recollection. The two states of awareness (remembering and knowing) are mutually excluding, but one can pass from the feelings of familiarity to active remembering through a process of active reconstruction.

Declarative or explicit memory is the ability to consciously recall items, facts and experiences, and then to express the recalled memories in many ways, more often by verbal language; *procedural* memory is inaccessible to conscious recall and appears through implicit measures of performance, such as increase in speed during performance of a previously learned skill, in priming, and in shift in the choice of recognised items.

Laboratory studies on skill acquisition have demonstrated *permanence and durability* of the learning of foreign language after very long periods (12 to 50 years), and procedural reinstatement of various abilities: motoric, perceptual skills such as swimming, playing tennis or piano, cognitive abilities such as speaking a foreign language or solving multiplication problems show a long durability if they are tested by *procedural reinstatement*; that is, the subjects who were given extensive training in some skill, at a retest after long periods are able to reinstate the sequence of operations, the procedures already acquired in precedent experience, although in a very distant time; the new test situation functions as a cue for retrieval, if it contains the same components of the original learning experience and if the rehearsal context is similar.

These observations are used by many authors in the discussion about modal models of memory. The early model, which descends from James' distinction between primary and secondary memory (1890)[3] receives a first scientific description by Atkinson and Shiffrin (1968)[4] who add sensory memory to the primary and secondary system; it receives a new formulation by Craik and Lockart[5] which model two level of processing, one at perceptual level, the second at a semantic level, that requires deeper processing but produces longer permanence (1972).

A modern variety is the Search Associative Model (SAM) by Raajmakers and coworkers,[6] who note that the first model of STM and LTM *stores* by Atkinson and Shiffrin becomes a two-processes *system* in the Craik-Lockart version, but has the flaw that it concerns only the encoding phase, neglecting the retrieval phase.

Raajmakers tries to fill this gap by his SAM, which assumes that during recall, stored information is represented in *memory images*, whose associative strength depends on the length of time during the acquisition, on the coding process during the Short Term Storage, on the overlap of the cue information and the information stored. The process of activating information is basically the same in recall and recognition: the overall activation started by the probe cues induces a *familiarity value* that is used in the manner of signal detection theory to determine the probability of recognition. This one is influenced not only by the strength of association, but also by the context, because changes in this context may determine the speed and accuracy of recognition response.

The model provides examples of predictions for free-recall, pair associate recall, interference effects, forgetting and various recognition phenomena, and the authors claim that the model is a powerful tool for analysing many memory experiments: but they admit that it cannot explain *retrieval from semantic memory* (which is not stored as image, and it is more context free) and other well established phenomena as *implicit memory*.

They suggest that the two store models should not be abandoned, but modified in a model that conceives episodic-semantic memory as a *continuum*, of which working memory and LTM (permanent store) are the extreme poles.

The distinction of STM and LTM is supported by many studies on amnesic patients who show selective deficits related to lesions in different brain areas: for instance ippocampal formation damage leads to severe amnesia in LTM, while does not diminishes performance in STM tasks like digit span or spatial memory. Others patients have deficit in STM memory such as free recall but show intact ability in LTM.

Baddeley (1986)[7] offers a model of working memory as a temporary storage and manipulation of information, with two mechanisms, the *visuo-spatial sketch pad*, the *articulatory loop* and a *central executive* which coordinates the two slave systems. The phonological loop seems to play an important role in L.T. phonological learning, the central executive is an attentional control system which selects and integrates information from different sources (it is severely disrupted in Alzheimer patients, more than the slave systems). In many amnesics there is dissociation not only between STM tasks and LTM tasks, but also between the last part of a list and the others: this recency effect can be explained as an implicit learning which is reactivated more easily since more recent items are primed.

Baddeley and coworkers (1974)[8] present many experiments which support the concept of phonological loop as a store which maintains information by subvocal rehearsal for a very short time, but this mechanism cannot explain sentence understanding, which surely is based on a conceptual representation: for instance when the sentences are presented in very rapid rate (faster than 3 words/sec.), the successive recall often is near perfect, or substitutes words by close synonymous, showing that the sentences are parsed and understood immediately; in other experiments ambiguous misspelled words are replaced with the correct words, suggesting that the sentences have been conceptually processed, and that conceptual representation derive from an activation of LTM despite the rapid rate of presentation.

In the same line Ericsson and Kintsch (1995)[9] present evidence that chess experts can increase their memory performance of spatial patterns drawing on precedent knowledge by a system of retrieval which permits to increase the capacity of storage in working memory using the information from LTM.

The same process appears in reading a text, when the subjects are familiar with the domain: expert readers have a superior and faster comprehension and recall, because while they are reading the new text they use material previously stored in order to encode and integrate the information.

This process is coherent with the model of LT-working memory, according which the active involvement of LTM in skilled performance suggests that STM is not a separate process, but *the activated portion of LTM on which the attention is temporary focussed*;[10,11] Baddeley too, in his most recent papers (1996),[12] emphasises the functions of the central executive which is quite similar to the supervisor attentional system by Norman and Shallice (1980);[13] in Italy Cornoldi (1995),[14] proposes an attentional continuum and Morra (1998),[15] suggests to abandon the concept of stores. Near all the psycholinguistic studies on lexical access show that the comprehension of ambiguous or fragmented words is related to the semantic context and might be due to an activation of LTM. We can see that in many types of experimental research the models of activation are related to a central control of attention which is stimulated by cues associated to prior experience.

Recent research in neuropsychology and in neurobiology offers an initial (perhaps still provisional) solution to the problem presented by the proliferation of models produced by cognitive psychology: in fact, as Roediger (1993)[16] observes, successive refinements of the dual SAM and Baddeley models have lead to increase hypothetical constructs which often are realised in nice elegant computer programs, but do not correspond to empirical observed phenomena: formal models are precise, definite, but very often are tested by some particular and very specific experimental paradigms, while they do not consider every day memory behaviour, which includes not only working memory, but many types of LTM, which is more difficult to test by carefully designed and controlled experimental situations. This area, that may be conceived as a critical crossing-point, but in the same time as a challenge and as a resource, has presented a great richness of new contributions from a series of researches in the field of every-day memory stimulated by the provocative criticism of Neisser (1982,[17] 1988[18]): to quote only the most important, autobiographical memory through the life-span, memory for personal and public events, the role of memory in problem solving and in lexical processing, eyewitness, remembering faces and names, and, finally, the study of aging.

MEMORY MECHANISMS

New advances have been offered by cognitive neuropsychology, starting from 1985 by the first investigation of implicit memory,[19,20] which focus the attention on the necessity to considerate not only conscious memory (intentional retrieval of information from a specific prior episode) but also unconscious memory, which does not implies conscious recollection of specific episodes.

The study of dissociations in brain damaged patients has given a new prospective approach and new directions, but sometimes has gone too far in postulating a number of memory systems or stores according to various dissociations due to subjects variables (brain damage or age) and experimental variables (priming tasks, methods of testing, extension or type of information retained, different forms of knowledge): so we have assisted to increasing fragmentation of mind into little components or modules - a new one for each new dissociation (Roediger, 1993).[16]

Schacter, who has given the first articulated treatment of this problem since 1985,[21] 1987, observes that implicit tasks such of as stem-completion and word identification, depend on *data driven* processing (i.e. *"bottom up"* processing, driven by perceptual features of materials) while explicit tasks, such as recall and recognition, are *conceptually driven*, because they are based on "*top-down*" processing actively initiated by the subject.

But the numerous proofs of the existence of dissociations between these processing are not proof of different systems: they are a necessary, but not sufficient condition, and the demonstration should come from neuroscience which should decide if the difference is due to different processes or to different systems.

In fact neurophysiology studies suggest the distinction between a *perceptual representation system* and a *semantic system*, more than between implicit and explicit memory.

The first one seems to be controlled by the right, the second one by the left hemisphere, which mediates the acquisition of language. Experiments on perceptual-structural access to words or objects show that the retrieval of the visual form of words and objects can be dissociated from semantic memory, suggesting that the perceptual processing is based on an independent system. Other amnesic patients show severe deficit in declarative memory, although they maintain learning of perceptuo-motor skills. In general, priming data driven of novel material may be intact while declarative memory for facts and events is severely impaired.

Squire and coworkers (1993)[22] have tried to review psychological research on multiple systems of memory comparing with data from neuroscience on the brain system.

They suggest that neurological base are different, since declarative (explicit) memories depend on neo-cortex, and are accessible to conscious recollections, while non declarative (or implicit) memory is an heterogeneous collection of memory abilities (skill learning, habits, conditioning, priming) which depends on different anatomical structures.

They support the distinction of STM and LTM comparing the dissociation between the two systems in different type of amnesic and in animals with experimental lesions: amnesic patients have severe deficit in conscious recollection and recognition of events and person (declarative memory), but preserve some forms of learning skills (mótor, perceptual and cognitive), of developing simple conditioned response and habits which can guide behaviour.

There is convergent evidence that limbic diencephalic structures with neocortex provide the basis for conscious recollection (declarative memory) while neo-striatum and cerebellum preserve skills, habits, permit priming and conditioning: in fact "*they provide dispositions, habit and preference that are inaccessible to conscious recollection, but are shaped by past events, influence our behaviour and are part of who we are*".[22]

More recently Eichenbaum (1997)[23] analyses the basic mechanisms of declarative memory arguing that hyppocampus supports the encoding of relational representations: spatial and temporal configurations of visual, auditory and spatial discriminations, matching objects, recognising spatial and temporal positions. Declarative memory is supported by cortical-hyppocampal circuit, which possibly has evolved as a mechanism for learning relations between cues, which is useful to survival.

Researches on animals behaviour in natural situations suggest that the learning characteristic is *plasticity*, which permits functional adaptation to tasks of different importance for survival: so the brain circuits evolved in specialised systems for spatial behaviour, singing, food storing, generating multiple forms of memory to face specific ecological problems. In humans the systems have become more specialised by the development of language and symbols.

These specialised systems originally were not accessible to conscious recollection, were rigid and specific, while through the natural evolution they developed complex and active organisation of experience which can be used with flexibility not only to recognise, but also to recall, to make inferences, to solve problems in novel situations.

This evolutionary view of procedural and declarative memory is confirmed by a new series of researches in the field of every day behaviour, that concern especially LTM, such as memory of autobiographical and public events, eyewitness, remembering of faces and names, the role of expertise in memory tasks, learning and problem solving, creative thinking.

RECOLLECTIVE MEMORY IN NATURAL CONTEXTS

Since the field of knowledge representation and organisation is too vast, I'll mention only two examples: *organisation of events* and *remembering of face and names*.

The study of autobiographical memories, started only since 1985 by Rubin,[24] has been rapidly increasing by different techniques which have shown that organisation of personal memories is not a fixed structure, but can change according to the method of retrieval: free recall, context-cued recall, recall according to temporal or spatial cues, starting from infancy, childhood to the present time or backward to precedent age periods, produce very different organisations, that show flexible reconstruction and multiple ways of activation - sometimes periods, sometimes single events, sometimes emotional states- whereas the only common feature is the *organisation around the concept of self* (which is congruent with the view of an adaptive evolution system).

Some researches on memory for events claim that events are organised together according activity, with subclusters of participants and locations locally embedded: but other experiments contradict this view, proving that people can cross-classify events into multiple organisations, and they can pivot between clusters by switching organisations:[25] clusters can be made according other events characteristics, such as people, locations, period of time.

Free recall, cued recall, retrieval of autobiographical memories can produce in the same person different organisations, which can shift, move and change in very flexible ways, showing the constructive and dynamic character of declarative memory that is regulated by frontal cortex.

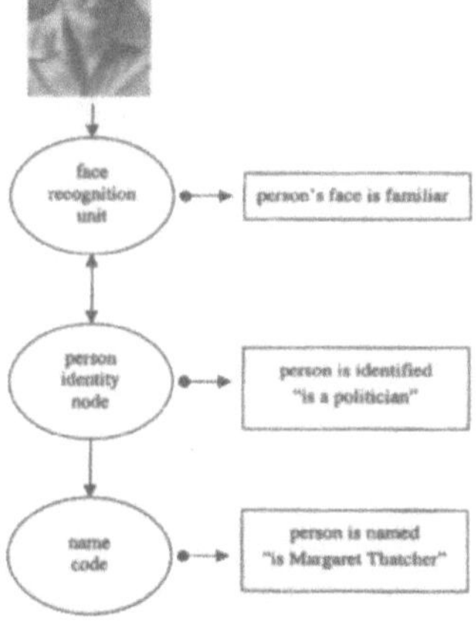

Figure 1. Memory for proper names: basic components of Bruce and Young's sequential stage model of face naming. (Modified from V. Bruce and A. Young, Understanding Face Recognition, *British Journal of Psychology*, 77: 305-327, 1986)[26]

If we examine another very rich set of experiments on the false memories we can observe the presence of many examples of unconscious, or implicit memory: we only briefly mention the case of *eyewitness*, in which very often people produce false memories which are unconsciously influenced by prior experiences, but we want to focus the attention to the well known phenomena of "*Tip Of the Tongue*" (TOT) and on the recall of proper names.

In both cases the subject is sure that he knows the word or the person name, but he cannot retrieve it: the difficulty in recalling the names of people that subject recognises presents an interesting analogy with other examples of the difference between knowing and remembering, since usually we recognise the profession, or the social identity of the person, while we are unable to retrieve the name, and in the same time if we try a wrong name, we are sure that it is wrong, but we remember the format, the beginnings and the end.

Proper names are more difficult to remember than other data on a person, like occupation, social status, role played in a film or a book. The difficulty of the retrieval processes appears in the *TOT* phenomenon, in which the subject cannot produce the right word, although he knows that he knows it.

This type of deficit increases during aging, but it is present also in states of fatigue or stress. In the case of aphasic patients it offers a strange variety of selective deficit, that is in some patients the impairment is limited to very specific categories, such as fruits, vegetables, objects, names of cities...

Again, it is impossible to imagine that each type of deficit is mediated by specific memory system or process; the most plausible set of hypothesis on person recognition are the models based on the difference between a semantic and a perceptual system.

Figure 1 shows one of the first models proposed by Bruce and Young (1986),[26] that describes a sequential process by three functional components: the first *(Face recognition unit)* is activated by some perceptual cue, the second retrieves semantic information from biographical, social or physical context *(Person identification unit)* the third is the access to the proper name.

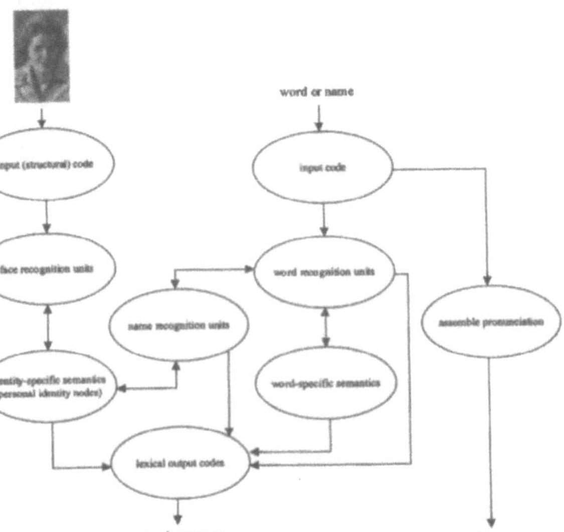

Figure 2. An interactive activation model: Valentine et al.'s functional model of face, name, and word recognition. (Modified from T. Valentine, S. Bredart, R. Lawson, and G. Ward, What's in a name? Access to information from people's name, *European Journ. Cognit. Psychol.*, 3, pp. 147-176, 1991)[27]

Mistakes may come from each stage, but they are more frequent in the last one because proper names are meaningless and arbitrary, have fewer links that semantic information, and receive poorer activation; so they are more affected by traumas, by aging and fatigue.

The sequential stage model of face naming has been recently revised and modified in various forms which take into account phenomena of language perception and production, memory and aging, types of representation: we mention only the recent reformulation of an interactive activation model of person recognition[28] because it supports our argument that it is not necessary to postulate different systems (see Fig. 2): in fact the processing units can be activated from different points and the process can go from recognition of face to professional activity, or vice versa, the name can be retrieved by association with the face while the semantic information is wrong (for instance, a political man is defined as an actor or a teacher or generically as a "bad person").

At the present moment the various experiments on remembering and knowing seem to suggest in some cases a difference between perceptual and semantic systems, between procedural and declarative memories, but in many others they can be better explained by a different level of activation of a unitary system which can be accessed from different points related to the input code and to previous personal experiences.

REFERENCES

1. J.M.Gardiner and R.I. Java, Recognizing and remembering, in: *Theories of memory*, A.F. Collins, S.E. Gathercole, M.A. Conway and P.E. Morris eds., Lawrence Erlbaum Associates, Hillsdale, NJ (1993).
2. E. Tulving, Haw many memory systems are there?, *American Psychologist* 40: 385-398 (1985).
3. W. James, *The Principles of Psychology*, Henry Holt, New York (1890).
4. R.C. Atkinson and R.M. Shiffrin, Human memory: a proposed system and its control processes, in: *The Psychology of Learning and Motivation: Advances in Research and Theory*, K.W. Spence and J.T. Spence eds., Academic, New York, 2: 89-195 (1968).
5. F.I.M. Craik and R.S. Lockart, Levels of processing. A framework for Memory Research, *Journal of Verbal Learning and Verbal Behavior* 11: 671-684 (1972).
6. J.G.W. Raajmarkers, The story of the Two Store Model of Memory: Past Criticisms, Current Status and Future Directions, in: *Attention and Performance*, XIV, D.E. Meyer and S. Kornblum eds., MIT Press, Cambridge, MA, 467-488 (1993).
7. A.D. Baddeley, R. Logie, S. Bressi and S. Della Sala, Dementia and working memory, *Quartely Journal of Experimental Psychology*, 38A: 603-618 (1986).
8. A.D. Baddeley and G.J. Hitch, Working memory, in:*The Psychology of Learning and Motivation: Advances in Research and Theory*, G.H. Bower ed., Academic, New York, 8: 47-89 (1974).
9. K.A. Ericsson and W. Kintsch, Long-term working memory, *Psychological Review*, 102: 211-245 (1995).
10. J.R. Anderson, *The Architecture of Cognition*, Harvard University Press, Cambridge, MA (1983).
11. N. Cowan, Activation, attention, and short-term memory, *Memory and Cognition* , 21: 162-167 (1993).
12. A.D. Baddeley, Exploring the central executive, *Quarterly Journal of Experimental Psychology*, 49A:5–28 (1996).
13. D.A. Norman and T. Shallice, Attention to action: Willed and automatic control of behaviour. Tecnical Report n. 99, Center for Human Information Processing. Reprinted (1986), in: *Consciousness and self-regulation*, vol. IV, R.J. Davidson, G.E. Schawartz and D. Shapiro eds., Plenum Press, New York, 1–18 (1980).
14. C. Cornoldi, La memoria di lavoro visuo-spaziale, in: *Le immagini mentali*, F. Marucci ed., La Nuova Italia, Firenze, (1995).
15. S. Morra, Magazzini di memoria? Pronti per l'oblio!, *Giornale Italiano di psicologia*, XXV, 4: 695–730 (1998).
16. H.L. Roediger III, Learning and Memory: Progress and Challenge, in: *Attention and performance*, XIV, D.E. Meyer and S. Kornblum eds., MIT Press, Cambridge, MA, 509-528 (1993).
17. U. Neisser, *Memory observed: Remembering in natural context*, W.H. Freeman and Co., S. Francisco (1982).
18. U. Neisser, Time present and Time past, in: *Practical Aspects of Memory: current Research and Issues*, M.M. Grunenberg, P.M. Morris and R.N. Sykes eds., Wiley, New York, 2: 545-560 (1988).

19. P.R. Graf and D.L. Schacter, Implicit and explicit memory for new associations in normal and amnesic subjects, *Journal of Experimental Psychology: Learning, Memory and Cognition* 11: 501-518 (1985).
20 D.L. Schacter, Implicit memory: history and current states, *Journal of Experimental Psychology: Learning, Memory and Cognition*, 13, 501-518 (1987).
21. D.L. Schacter, Multiple forms of memory in human and animals, in: *Memory Systems of the Brain*, N.M. Weinberger, J.L. McGaugh, G. Lynch eds., Guilford, New York, 351-380 (1985).
22. L.R. Squire, B. Knowlton and G. Musen, The structure and Organization of memory, *Annual Review of Psychology*, 44: 453-495 (1993).
23. H. Eichenbaum, Declarative memory: Insights from Cognitive Neurobiology, *Annual Review of Psychology* , 48: 547-572 (1997).
24. D.C. Rubin, *Autobiographical Memory*, Cambridge University Press, Cambridge, MA (1986).
25. J.S. Lancaster and L.W. Barsalou, Multiple Organizations of Events in Memory, *Memory*, 5 (5): 569–599 (1997).
26. V. Bruce and A. Young, Understanding Face Recognition, *British Journal of Psychology*, 77: 305-327 (1986).
27. T. Valentine, S. Bredart, R. Lawson, and G. Ward, What's in a name? Access to information from people's name, *European Journ. Cognit. Psychol.*, 3, pp. 147-176 (1991)
28. A.M. Burton and V. Bruce, Naming Faces and Naming Names: Exploring an Interactive Activation Model of Pearson Recognition, in *Memory* , 4: 457-80, Special Issue edited by G. Cohen and D.M. Burke (1993).

PANEL SUMMARY: **SYMBOLISM AND CONNECTIONISM PARADIGMS**

Panellists: Lorenzo Magnani (*chairperson*),[1] Antonio Chella,[2] Luciano da Fontoura Costa[3]

[1]Computational Philosophy Laboratory, Department of Philosophy
University of Pavia I-27100 Pavia, Italy
e-mail:lorenzo@philos.unipv.it
[2]Department of Electrical Engineering, University of Palermo
Viale delle Scienze, I-90128 Palermo, Italy
e-mail: chella@unipa.it
[3]DFI-IFSC, Universidade de Sao Paulo
Caixa Postal 369, Sao Carlos, SP 13560-970 Brazil
e-mail: luciano@ultra3000.ifsc.sc.usp.br

INTRODUCTION

The aim of this chapter is to report the panel discussion on symbolism and connectionism paradigms. In particular, the following hot point are analysed:

- what cognitive phenomena are most difficult for connectionists to explain?
- what cognitive phenomena are most naturally explained in connectionist terms?
- is symbolic deduction a central kind of human thinking? How do people make deductions?
- is nondeductive reasoning done in accord with the laws of probability?
- what areas of knowledge do you have that are easily described in terms of symbolic rules?
- concepts reduced to rules, concepts reduced to networks;
- symbolic and connectionist mechanisms of analogy;
- planning, decision, explanation, learning, language, in front of the symbolic/connectionist dichotomy.

EVALUATING COMPUTATIONAL-REPRESENTATIONAL APPROACHES
(*L. Magnani*)

In his last book Paul Thagard[1,p.10] summarises the *central hypothesis of cognitive science*: «Thinking can best be understood in terms of representational structures in the mind and computational procedures that operate on those structures. Although there is much disagreement about the nature of the representations and computations that constitute thinking, the central hypothesis is general enough to encompass the current range of thinking in cognitive science, including connectionist theories». This approach is called *Computational-Representational Understanding of Mind.* Hence, representation and computation are viewed as the two major fields of research and logic, rules, concepts (frames, semantic networks, and so on), analogies, images, and connections (artificial neural networks) are considered the most important ways of representation and computation.

Thagard[1,p.15] gives us an analysis of all these approaches (logic, rule, concepts, analogies, images, and connections) in terms of five complex criteria:

1. representational power
2. computational power
    a. problem solving
        i. planning
        ii. decision
        iii. explanation
    b. learning
    c. language
3. psychological plausibility
4. neurological plausibility
5. practical applicability
    a. education
    b. design
    c. intelligent systems

I propose to adopt this framework because it is very useful to correctly pose the problem of the dichotomy symbolism/connectionism from a theoretical point of view.*

The first criterion, *representational power*, deals with the quantity of information a particular kind of representation can express. Different ways of representation vary very greatly in representational power. Moreover, a computational approach to mental representations surely has to account for three important kinds of high-level reasoning: *problems solving* (planning, decision making, and explanation), learning, and language. We can evaluate the *computational power* of a particular representation in terms of how it regards these kinds of reasoning. The ability to *learn* from experience is also an important part of intelligent reasoning, and the capability to explain the use of *language* is a desirable aspect of a general cognitive theory.

The *psychological plausibility* of a theory of mental representation aims to account for quantitative and qualitative results of psychological experiments concerning certain mental capabilities of humans. A theory of mental representation should be also consistent with the results of neuroscientific experiments.

Finally, the *practical applicability* concerns the effective results in the different areas of education (for instance, how to teach better), design (for instance, new computer interfaces), and intelligent systems (expert and knowledge-based systems - rule-based, case-based, connectionist - in many fields of artificial intelligence).

Cognitive Tasks and Representations

The cognitive accomplishments made in cognitive science by the approaches in terms

---

*Let us remember that a second section of the book is devoted to the so-called «challenges to cognitive science», considered as different ways to object to the Computational-Representational Understanding of Mind, and deals with the problem of the role of emotion and of consciousness in human thinking, of physical and social environments, of dynamic systems and mathematical knowledge (the mind is a dynamic system, not a computational one, mathematical results show that human thinking is not computational in the usual sense, but may be it operates as a quantum computer).

of logic, concepts, rules, analogies, and imagery, are well-known. We have very large computational systems that employ the resources of *logic*,[2] the most ancient way of representing knowledge, sometimes with the help of probabilistic theory.[3] *Concepts* (in terms of frames, scripts, semantic networks),[4,5] and *rules* (as heuristic search and rule-based chunking)[6-10] constitute the background of many other systems able to model many cognitive aspects of language and problem solving. *Analogical thinking* (also called case-based reasoning)[11] and *imagery*[12-14] are subject to very new and interesting research that involve psychological experiments, computational programs, and neurological examinations.

Hence, concepts, analogies, and images are very useful to provide many organised clusters of representations, rules and propositions in logic are useful because of their verbal expressiveness, and images, imagery and connectionism because of their sensory richness. What exactly about connectionism? Let us devote more time to illustrate some details concerning the connectionist approach.

If we consider the two well-known classes of neural networks, *local* and *distributed*, in the first class the neuronlike structures are given an identifiable interpretation in terms of concepts or propositions, while in the second one the distributed representations in networks learn how to represent concepts or propositions in complex intricate ways that distribute meaning over clusters of neuronlike structures: both local and distributed representations can be used to perform the so-called *parallel constraint satisfaction*.

It is well-known the first models of parallel constraint satisfaction were first developed for *computer vision*;[15-17] it is also well-known that links between units in a distributed network are adequate to represent simple associations but lack the representational power to grasp more complicated kinds of rules such as that ones containing some uses of universal quantifiers and complex logical relations. Nevertheless, neural networks offer more flexibility in grasping a broader range of sensory* experience, for instance they make possible many more tastes and aromas than we can say in words.[18]

Constructing a plan is naturally understood in terms of rules and analogies, but *decision* among plans is easier to represent using parallel constraint satisfaction[19,20] because of the fact we can interpret the facility relations as positive internal constraints, and the negative internal constraints as incompatibility relations, when two actions or goals cannot be performed or satisfied together. We have also to remember that connectionist networks can implement simple types of ruled-based systems[21] but also the «mental leaps» of the *analogical reasoning*.[22,11]

In its turn *explanation* can be seen as activation of prototypes represented by distributed networks;[23] on this perspective, *inference to the best explanation* can be viewed as the activation of the suitable prototype. For example, the theory of explanatory coherence,[24,25] implemented in a connectionist local neural network, is able to activate the hypotheses (or propositions) that better explain (that is, «they have more explanatory coherence») some clusters of evidence.

Adding new units and changing the weights - by training - on the links between units to a connectionist network are the main ways to perform *learning* tasks:[26] the so-called feedforward networks trained by backpropagation techniques have had many successful applications.

Finally, McClelland and Rumelhart have also illustrated that word recognition can be considered from the point of view of a parallel constraint satisfaction, and many other connectionist tools have implemented other cognitive linguistic tasks: disambiguation, speech perception, learning verbs by children (and producing errors in that learning).

We can conclude that the psychological plausibility of connectionist models is very high: from the phenomena of word perception[17] to discourse comprehension,[27,28] from

*Remember that the richness of sensory experience is also provided by the cognitive approach in terms of images or imagery.

analogical mapping[11] to visual word recognition.[29] Connectionist weight adjustment is a very powerful mechanism of learning language, even if we do not dispose of a unified connectionist theory of language production and comprehension. If compared with the approaches in terms of logic, rules, and concepts, the connectionism has an obvious neurological plausibility: we can say that most connectionist models are a very rough approximation to the behaviour or the real (human) neurons.*

At present cognitive science does not show a unified theory able to explain a full range of psychological phenomena,** similar to the genetic theory in biology or quantum theory in physics: all the approaches we have illustrated above give rise to different autonomous points of view concerning the «mind». We can say that the different approaches stress different representational and inferential aspects of mind, have different psychological and neurological plausibility, and many original applications in education, design, expert and knowledge-based systems.

The consequence is that an eclectic attitude is imperative, not only from the epistemological point of view, but also in the activity of the knowledge engineer. Some researchers insistently refer to the need of an hybrid approach, an integration of symbolic and neural representation and computation, in areas such as linguistic theory,[30] natural language processing[31] and expert systems,[32] where symbolic approaches have been prevalent in the past. The opportunity of using neural techniques is usually suggested by the fact that neural networks can reduce well-known shortcomings of symbolic problem solvers: for example, brittleness caused by incomplete data, no increase in performance with experience, time consuming knowledge acquisition, sensory processing needed by data acquisition, exploitation of sensory information in the training phase to provide pictorial non-symbolic forms of explanation.[32,33]

SYMBOLIC AND SUBSYMBOLIC REPRESENTATIONS: AN ARTIFICIAL VISION FRAMEWORK (*A. Chella*)

A case study of the interplay between the symbolic and subsymbolic representation, a framework for visual knowledge representation at different levels of abstraction is here discussed. The framework is aimed to integrate in a principled way the approaches developed within the artificial vision community with the propositional systems developed within symbolic knowledge representation in AI.[34,35]

The computer vision community approached the problem of representation of dynamic scenes mainly in the terms of their 3D reconstruction and of the recovery of motion parameters even in the presence of noise and occlusions. On the other side, the AI community developed rich and expressive systems for representation of time, actions and, in general, of dynamic situations.

Nevertheless, these two traditions evolved separately and concentrated on different kinds of problems.[36] On the one side, the computer vision researchers implicitly assumed that the problem of visual representation ends with the 3D reconstruction of moving scenes. On the other side, AI community usually did not face the problem of grounding the representations on the data coming from sensors.

The existing attempts directed towards an integration of visual perception with symbolic representations are mainly oriented to natural language interpretation, with particular attention to the field of man-machine interaction. They face only in a marginal

---

*Real neurons are of course more complex than the units of the artificial networks, they have many neurotransmitters that involve chemical links, so the brain has to be considered not only in electrical terms, but also in chemical ones, neurons show changes in synaptic and nonsynaptic properties that are not represented in artificial neural networks; moreover, backpropagation learning does not correspond to the processes observed in the brain.

**Newell[8] thinks to a unified theory of cognition in terms of rules.

way the general aspects of knowledge representation. Furthermore, the existing proposals often concern very specific domains of application.

On the contrary, our aim is to propose a principled integration of the approaches of artificial vision and of symbolic knowledge representation. We assume that this integration requires the introduction of a missing link between these two kinds of representation. The role of such a link is played by the notion of *conceptual space*. We borrow the notion of conceptual space from Gärdenfors.[37] A conceptual space is a representation where information is characterised in terms of a metric space defined by a number of *cognitive* dimensions. These dimensions are independent from any specific representation language. The conceptual space acts as an intermediate representation between subconceptual knowledge (i.e., knowledge that is not yet conceptually categorised), and symbolically organised knowledge.

According to this perspective, the proposed architecture is organised in three computational areas:

- the *subconceptual* area;
- the *conceptual* area;
- the *linguistic* area.

These areas must not be interpreted as a hierarchy of levels of higher abstraction; rather, they are concurrent computational components working together on different commitments.

The *subconceptual* area is concerned with the low level processing of perceptual data coming from the sensors. The term subconceptual suggests that information is not yet organised in terms of conceptual structures and categories. In the *linguistic* area representation and processing are based on a logic oriented formalism. In the *conceptual* area, the data coming form the subconceptual area, are organised in conceptual categories; however, they are still independent from any linguistic description.

These three areas are highly interconnected. There is no privileged direction in the communications among them: some computations are strictly bottom up with data flowing from the subconceptual up to the linguistic area through the conceptual area; other computations combine top-down with bottom up processing.

The mapping between the conceptual and the linguistic areas is based on a *focus of attention* mechanism, implemented by recurrent neural networks.

We refer to a scenario in which two persons stand at the two sides of a table with some objects on it. The persons may perform several actions: they may push a block; they may grasp it; they may lay down the block on the table, and so on. Images of the scene are acquired by a colour CCD camera and processed by a 3D computer vision system, which recovers the moving shapes in the scene in terms of geometric and motion parameters of simple 3D primitives as cylinders, boxes and spheres. This experimental setup is simplified enough for our purposes; however, it is sufficient to allow many inferences and to build rich representations.[38] It should be noted that the proposed framework is not limited to this special purpose domain, but it may be applied in all cases in which abstract reasoning on sensory data is needed.

Conceptual Area Representation

The representation in the conceptual area is based on the previously cited notion of conceptual space that provides a principled way for relating a low level, unstructured, representation of data with a high level, linguistic formalism. Formally, a conceptual space is a metric space whose dimensions are strictly related to the quantities processed in the subconceptual area (e.g., sensory related quantities). Examples of possible dimensions could be color, pitch, mass, spatial coordinates, and so on; they do not depend from any specific linguistic description. In this sense, conceptual spaces are prior to symbolic characterisation of cognitive phenomena.

We adopt the term *knoxel*[34] to denote a point in the conceptual space. A knoxel is an epistemological primitive element at the considered level of analysis. We assume that, in the case of static scenes, a knoxel coincides with a 3D primitive shape, as a cylinder or a box.

The entities represented at the linguistic level usually do not correspond to single knoxels. We therefore assume that *complex entities* correspond to set of knoxels. For example, in the case of a static scene, a complex shape corresponds to the set of knoxels of its simple constituents.

The expressions at the linguistic level are interpreted on structures in the conceptual space. This interpretation is in a certain sense similar to the usual interpretation of formal languages according to model theoretic semantics. However, conceptual spaces offer the advantage of a richer semantic structure with respect to usual set theoretic techniques. In a conceptual space a metric is defined, according to which similar entities correspond to neighbouring knoxels in the space. Moreover, convex clusters of knoxels are good candidates for the interpretation of the linguistic symbols expressing natural categories.

To account for dynamic scenes, we choose to adopt an *intrinsically dynamic* conceptual space. It has been hypothesized[39] that simple motions are perceived in their wholeness and not as a sequence of single static frames. Therefore, we model the conceptual space so that every knoxel corresponds to a simple motion of a 3D shape primitive. In other words, we assume that simple moving objects are our perceptual primitive entities. In particular, in the experimental setup the 3D primitives are cylinders, spheres and boxes.

For example, consider a person that moves his arm up and down. The first part of the trajectory of the forearm is a simple motion corresponding to a knoxel *k1*. When the arm has reached its vertical position, an instantaneous rest state occurs; the second part of the trajectory of the forearm is again a simple motion corresponding to a new knoxel *k1'*. The same holds for the upper arm: the first part of its trajectory corresponds to a knoxel *k2* the second part, after the instantaneous rest state, corresponds to a new knoxel *k2'*.

In this intrinsically dynamic conceptual space, a knoxel corresponds to a generalised simple motion of a 3D primitive shape. The motion is generalised in the sense that it can be decomposed in a set of motions, each one associated with a degree of freedom of the geometric primitive. In its turn, these motions corresponding to the degrees of freedom of the primitive, can be viewed as the result of the *superimposition* of elementary basic motions. In this way, it is possible to individuate a set of basis functions that make the space a functional space. In the specific case study, the motions corresponding to each degree of freedom can be viewed as the result of the superimposition of the first low frequencies harmonics in the context of the well-known Fourier analysis. The choice to consider only the first low-frequencies is due to the fact that the motions occurring in the scenario of persons moving their arms and moving objects are generally smooth.

The *composite_motion* is the motion of a composite object (i.e., an object approximated by more than one atomic 3D shape). A *composite_motion* is represented by the set of knoxels corresponding to the motion of its components. In *composite_motion*, the simple motions occur simultaneously, i.e., they correspond to a single configuration of knoxels in the conceptual space.

In order to consider a composition of several (simple or composite) motions arranged according to some temporal relation (e.g., a temporal sequence), we introduce the notion of *action.*[40] An *action* corresponds to a sequence of different configurations of knoxels in the conceptual space. We assume that the configurations within an action are separated by instantaneous events. In the transition between two subsequent configurations, a *scattering* of at least one knoxel occurs. This corresponds to a discontinuity in time that is associated to an instantaneous event.

Linguistic Area Representation

Representation in the linguistic area is based on a high level, logic oriented formalism, in particular a hybrid formalism in the KL-ONE tradition.[41] An hybrid formalism in this sense is constituted by two different components: a *terminological* component for the description of concepts, and an *assertional* component, that stores information concerning a specific context. In the domain of dynamic scene representation, the terminological component contains the description of relevant concepts such as various types of motion, of action, and so on. The assertional component stores the assertions describing specific perceived situations.

Focus of Attention

The mapping between a symbolic representation in the linguistic area and a structure in the conceptual space is based on a focus of attention mechanism. A finite agent with bounded resources cannot carry out a one shot, exhaustive, and uniform analysis of acquired data within reasonable time constraints. Some of the acquired data are more relevant than others, and it should be a waste of time and of computational resources to detect true but useless details.[42] In the model, the focus of attention allows to select the relevant aspects of a perceived dynamic scene by sequentially scanning the knoxels in the conceptual space. The focus of attention is crucial in determining which assertions must be added to the linguistic knowledge base: not all true (but often useless) assertions are generated, but only those that are judged to be relevant on the basis of the attentive process.

The recognition of a certain kind of simple motion (a knoxel) will elicit the expectation of other related simultaneous motions in the scene (e.g., motions of other parts of a complex shape). In this case, the mechanism seeks for the corresponding knoxels in the current conceptual space configuration. We call this type of attention *synchronic* because it refers to a single configuration of knoxels.

The recognition of a certain configuration in the conceptual space could also elicit the expectation of change in the arrangement of the knoxels. In this case the mechanism seeks for corresponding knoxels in subsequent configurations of conceptual space. We call this attention *diachronic*, in the sense that it involves subsequent configurations of conceptual space.

We take into account two main sources of expectations playing a role in the attentive process. On the one side, expectations could be generated on the basis of the structural information stored in the symbolic knowledge base. For example, the stretching out of an arm is described as composed by two simple motions: the motion of the forearm and the motion of the upper arm. We call *linguistic* these expectations. As soon as either one of the two motions is recognised, the symbolic description elicit the expectation of the other.

On the other side, expectations could also be generated by purely associative, Hebbian mechanism. Suppose that many examples of scenes have been seen, where one agent offered an object to the other that took it. The system could learn to associate these kinds of motion: when one agent stretches his arm holding an object towards the other, this one is expected to grasp it. We call *associative* these expectations.

Note that both symbolic and associative mechanisms have a role in both synchronic and diachronic expectations.

Mapping by a Neural Network

In the present model, the mapping between the conceptual space and the linguistic area is implemented according to a connectionist approach, by means of recurrent neural network. In particular, each primitive motion and action concept in the linguistic area is associated with a suitable neural network of the Hopfield type with asymmetric delayed connections[43,44] which acts as a *predictive filter* on the sequences of knoxels corresponding to instances of the concept.

Let me consider the set of knoxels {*k1*, *k2*, ..., *km*} corresponding to an instance of a generic concept. When a knoxel of the concept, say *k1*, has been individuated in the conceptual area and it is presented as input to the network associated to the concept, the network generates as output another knoxel of the set, say *k2*. In this way, the network predicts the presence of *k2* in the conceptual space. The expectation is considered confirmed when the conceptual area individuates a knoxel *k** close enough to *k2*. If the expectation is confirmed, then the network receives as input *k2* and generates a new expected knoxel *k3* and so on. The network therefore recognises the configuration of knoxels of the associated concept according to a recognition and expectation loop.[38]

If the concept is a composite motion, then the sequences of knoxels will refer to the same configuration of conceptual space. If the concept is an action, the sequences of knoxels will refer to subsequent configurations of conceptual space. The first case is an example of synchronic attention, the second one is an example of diachronic attention.

Neural networks make it possible to avoid an exhaustive description of conceptual categories at the linguistic area: in some sense, prototype motions and actions «emerge» from the activity of an associative mechanism during the training phase based on examples. In addition, the measure of similarity between a prototype and a given motion or action is implicit in the behaviour of the network and is determined during the learning phase.

COMMENTS (*L. da Fontoura Costa*)

I would like to express my comments about some of the hot topics of the panel.

• What cognitive phenomena are most difficult for connectionists to explain?

• What cognitive phenomena are most naturally explained in connectionist terms?

• Planning, decision, explanation, learning, language, in front of the symbolic connectionist dichotomy.

Connectionist models, at least in the way they have been understood and considered, are best for explaining low level cognitive processes. Such processes are typically characterised by being local, involving relatively disaccoupled subprocesses, and being more parallel (in execution). In vision, for instance, connectionist models have been more successful in modelling edge enhancement and detection. One of the reasons for the relative success of connectionist models at this hierarchical level is due to the close relationship between the local operations (for instance convolution with a filter mask) and the immediate definition of the neural weights. As a matter of fact, any convolution-based filter can be directly mapped into neural networks.

The problems with the connectionist approach start to appear as we proceed further into the brain hierarchy, and the processes become more interdependent, global and sequential. Two important related issues are those involving representations and rules. How are these represented and processed in neural systems? How can be rules represented and extracted from neural networks? These are important questions which have motivated much debate in the related communities. But they seem to be far from properly answered.

• Is symbolic deduction a central kind of human thinking? How do people make deductions?

That it is important, there should be no doubt. After all, we are taught to think in that way from our early years. Our subconscious, and even conscious, high level processing are deeply influenced by the deductive approach. The question that remains is what are the other types of reasoning. Also, is there any sort of «natural» way of reasoning? Does such a natural way vary along the brain hierarchy?

• Is nondeductive reasoning done in accord with the laws of probability?

According to the statistical mechanics approach to neural networks, the brain could be explained in terms of statistics. If that is really so, there should be little wonder that nondeductive reasoning is controlled by probability laws. However, there is another important perspective that should not be overlooked, namely that regarding variational calculus. It is known from physics that nature seems always to act according to the least effort principle, from quantum mechanics to astrophysics. Thus, the probabilistic description should perhaps reflects the energy aspects of biological neural processing.

• What areas of knowledge do you have that are easily described in terms of symbolic rules?

• Symbolic and connectionist mechanisms of analogy.

• Planning, decision, explanation, learning, language, in front of the symbolic connectionist dichotomy.

Those which are less ambiguous and involve less subjective concepts, such as some laws (e.g. those regulating soccer and other games) and algorithms. Even so, the symbolic description of human concepts is not straightforward because of the subtleties and subjectiveness of our own language. There should be no doubt that exclusively symbolic approaches are only part of the solution to the problem of cognition. We need sound means for embedding statistical features and generalisations into such rules.

• Concepts reduced to rules, concepts reduced to networks.

Three important questions arise here:

(1) what sort of concepts can be reduced to rules?

(2) how can this be done?

(3) why should concepts be reduced to networks?

The first question is actually relative to the sort of concepts and rules we have. The involved concepts can only be properly represented by rules if these are capable of properly incorporating their subjective and statistical nature. But what about question (3)? Though it is inherent in biology, since neural systems are the means for information processing, it is not clear why we should necessarily map concepts into networks in artificial reasoning systems.

ACKNOWLEDGEMENTS

The authors would like to thank all the participants of the panel who proposed interesting ideas and suggestions.

REFERENCES

1. P. Thagard, *Mind. Introduction to Cognitive Science*, The MIT Press, Cambridge, MA (1996).
2. M. Genesereth and N. Nilsson, *Logical Foundations of Artificial Intelligence*, Morgan Kaufmann, Los Altos, CA (1987).
3. J. Pearl, *Probabilistic Reasoning in Intelligent Systems*, Morgan Kaufmann, San Mateo, CA (1988).
4. M. Minsky, A framework for representing knowledge, in: *The Psychology of Computer Vision*, P. Winston ed., McGraw-Hill, New York (1975).
5. R.C. Schank and R.P. Abelson, *Scripts, Plans, Goals, and Understanding, An Inquiry into Human Knowledge Structures*, Erlbaum, Hillsdale, NJ (1977).
6. A. Newell, C. Shaw, and H. Simon, Elements of a theory of human problem solving, *Psychological Review* 65, 151-166 (1958).
7. A. Newell and H. Simon, *Human Problem Solving*, Prentice-Hall, Englewood Cliffs, NJ (1972).
8. A. Newell, *Unified Theories of Cognition*, Harvard University Press, Cambridge, MA (1990).
9. J.R. Anderson, *Cognitive Science and its Implications*, Freeman, New York (1980).
10. J.R. Anderson, *Rules of the Mind*, Erlbaum, Hillsdale, NJ (1993).

11. K. Holyoak and P.Thagard, *Mental Leaps: Analogy in Creative Thought*, The MIT Press, Cambridge, MA (1995).
12. S.M. Kosslyn, *Image and Brain: The Resolution of the Imagery Debate*, The MIT Press, Cambridge, MA (1994).
13. S.M. Kosslyn and O. Koenig, *Wet Mind. The New Cognitive Neuroscience*, Free Press, New York (1992).
14. J.I. Glasgow and D. Papadias, Computational imagery, *Cognitive Science* 16, 355-394 (1992).
15. D. Marr and T. Poggio, Cooperative computation of stereo disparity, *Science* 194, 283-287 (1976).
16. J. Feldman, A connectionist model of visual memory, in: *Parallel Models of Associative Memory*, G. E. Hinton and J. A. Anderson eds., Erlbaum, Hillsdale, NJ (1981).
17. J.L. McClelland and D.E. Rumelhart, An interactive activation model of context effects in letter perception. Part I, An account of basic findings, *Psychological Review* 88, 375-407 (1981).
18. P.M. Churchland, *The Engine of Reason, the Seat of the Soul*, The MIT Press, Cambridge, MA (1995).
19. P. Thagard and E. Millgram, Inference to the best plan: a coherence theory of decision, in: *Goal-Driven Learning*, A. Ram and D. B. Leake eds., The MIT Press, Camdridge, MA (1995).
20. S.M. Mannes and W. Kintsch, Routine computing tasks: planning as understanding, *Cognitive Science* 15, 305-342 (1991).
21. D. Touretzky and G. Hinton, A distributed production system, *Cognitive Science* 12, 423-466 (1988).
22. G. Nelson, P. Thagard, and S. Hardy, Integrating analogies with rules and explanations, in: *Advances in Connectionism vol. II*, K. Holyoak and J. Barnden eds., Ablex, Norwood, NJ (1994).
23. P.M. Churchland, *A Neurocomputational Perspective*, The MIT Press, Cambridge, MA (1989).
24. P. Thagard, Explanatory coherence, *Behavioral and Brain Sciences* 12, 435-467 (1989).
25. P. Thagard, *Conceptual Revolutions*, Princeton University Press, Princeton (1992).
26. D.E. Rumelhart, J.L. McClelland, and PDP Research Group, *Parallel Distributed Processing: Explorations in the Microstructure of Cognition*, The MIT Press, Cambridge, MA (1986).
27. D.E. Rumelhart and J.L. McClelland., An interactive activation model of context effects in letter perception. Part 2, The contextual enhancement effect and some tests extensions of the model, *Psychological Review* 89, 60-94 (1982).
28. W. Kintsch, The role of knowledge in discourse comprehension: A construction-integration model, *Psychological Review* 95, 163-182 (1988).
29. M.S. Seidenberg and J.L. McClelland, A distributed, developmental model of word recognition and naming, *Psychological Review* 96, 523-568 (1989).
30. A. Prince and P. Smolenski, Optimality: from neural networks to universal grammar, *Science*, 275 1604-1610 (1997).
31. M.G. Dyer, Symbolic neuroengineering for natural language processing: a multilev approach, in: *Advances in Connectionist and Neural Computation Theory vol I*, J.A. Barnden and J.B. Pollack eds., Ablex, Norwood, NJ (1991).
32. L.R. Medsker, *Hybrid Neural Networks and Expert Systems*, Kluwer, Dordrecht (1994).
33. E. Burattini, M. De Gregorio, and G. Tamburrini, Hybrid expert systems. An approach combinig neural computation and rule-based reasoning, in: *Expert Systems Techniques and Applications*, C. D. Leontes ed., Gordon and Breach, London (1998).
34. A. Chella, M. Frixione, and S. Gaglio, A cognitive architecture for artificial vision, *Artificial Intelligence* 89, 73-111 (1997).
35. A. Chella, M. Frixione, and S. Gaglio, An architecture for autonomous agents exploiting conceptual representations, *Robotics and Autonomous Systems* 25(3-4), 231-240 (1998).
36. M. Brady and H. Hu, The mind of a robot, *Phil. Trans. R. Soc. Lond. A*, 349:15-28 (1994).
37. P. Gärdenfors, *Conceptual Space*, Lund University Cognitive Science (1998).
38. D. Marr and L. Vaina, Representation and recognition of the movements of shapes, *Proc. R. Soc. Lond. B* 214, 501-524 (1982).
39. C.R. Gallistel, *The Organization of Action: A New Syntesis*, Lawrence Erlbaum Associates, Hillsdale, NJ (1980).
40. J.F. Allen, Towards a general theory of action and time, *Artificial Intelligence* 23(2), 123-154 (1984).
41. R.J. Brachman and J.C. Schmoltze, An overview of the KL-ONE knowledge representation system, *Cognitive Science*, 9(2), 171-216 (1985).
42. D.H. Ballard, Cortical connections and parallel processing: structure and function, *Behavioral and Brain Sciences* 9, 67-120 (1986).
43. J.J. Hopfield, Neural networks and physical systems with emergent collective computational abilities, *Proc. Nat. Acad. Sci. USA* 79, 2554-2558 (1982).
44. D. Kleinfeld, Sequential state generation by model neural networks, *Proc. Nat. Acad. Sci. USA* 83, 9469--9473 (1986).

PRESENCE, TELEPRESENCE AND USABILITY

Stefano Levialdi

Pictorial Computing Laboratory
Dipartimento di Scienze dell'Informazione
Università di Roma, La Sapienza
Via Salaria 113, I-00198 Roma (Italy)

INTRODUCTION

This chapter introduces the concept of presence and telepresence, as commonly used in different research areas and applications, noting the subtle differences in meaning when referred to a different media. The ambiguity of such terms stems from the contexts onto which they are applied, on one hand, and on the different discipline which is observing the effects (on humans, on programs, on robots); in fact many areas are involved in explaining, applying and evaluating how humans feel in a different place from the one they are really living. Psychologists and cognitive scientists refer to the consequences of "being elsewhere" through a variety of media (television, the web, and other technologies) whilst computer scientists try to provide "transportation", generally through visual cues, so as to enable humans to perceive objects even if they are remotely placed. Finally, the chapter ends by recalling the recent concept of usability of information systems and showing its relevance for an effective implementation of telepresence in a recent project devoted to teleoperation.

THE TITANIC WRECKAGE: AN APPRAISAL EXPERIMENT

Let us start with an example which demonstrates the different "realities" which might be important to document and understand what happened the night during which an iceberg collided with the Titanic and which were the consequences of such a colossal accident on the North Atlantic Ocean. There are many different points of view when a documentation of an accident is required and the only real objects left are in the bottom of the ocean. At the beginning, a number of persons thought that the only way to have a direct view of the remains of the passenger ship was to pick whatever was left in the water and bring it to shore. When it was discovered that both the price of rescuing the remains of the ship and the danger of ruining such remains made the enterprise a very risky business, it was decided to leave things as they were. Similar effects could be obtained by submerging a telecamera under the water and viewing - at a distance - whatever was left of the ship or, alternatively,

Figure 1. A virtual reconstruction of the Titanic staircase descending towards the dancing room.

reconstructing - by means of virtual reality - the different dining rooms and saloons which made the Titanic so famous, allowing different views, panning, zooming, etc. One view of the Titanic staircase is shown in Figure 1.

Another approach, using enacted reality, could be done (in fact there was a musical comedy on this subject some years ago) like building a set at the cinema studios where a full reconstruction of the ship was performed and the sinking, as well as other representative moments of the wreckage, were filmed in the contemporary Titanic film by J. Cameron. We see here that a variety of "realities" are possible in which different technologies can be used each one producing a specific degree of realism, i.e. recreating the scene, the time and the actions of the characters involved.

PRESENCE

In many of today's activities, the concept of being somewhere else (where the action is on) is relevant to the completion of the task; for example in simulation where pilot training, video games and urban planning (to name a few) are all dependent on the quality of the reproduction of an invented reality. In video conferencing the illusion of being close is fundamental to enable a smooth conversation that will lead to decisions, to new policies, to different forms of collaborative work and management, etc. Entertainment is certainly another area in which virtual reality, IMAX films, sound as produced by recent Dolby-Surround marked systems, and home theatre improve as soon as the "presence" effect is produced by increasing the audio channels, by filling the air with high quality sounds. Finally, education with its recent boom in distance learning, web-based learning and interactive teaching are re-creating the schools, universities, institutions so fast that we have today over 1500 virtual universities that may be reached from anywhere in the world to obtain credits in a new teaching-learning style called "contract learning".

In short, the concept of presence may be described in terms of a mediated experience that humans may have in order to achieve a natural-immediate-direct-real perception. The Webster Dictionary (1997) states that presence stands for "being at a specified or understood place" whilst a person is present, according to the Concise Oxford Dictionary (1981) when the "when being in the place in question". These definitions, even if a little reductive, convey the principal component of the meaning. A recent definition by Kim et al.,[1] sees presence as the "person's perception of being at a specified or understood place" (borrowing from the Dictionary), but another, more thorough definition by Lombard et al.[2] considers presence as a "mediated experience allowing the perception of a different physical/virtual place" (from the real one, in which the person is physically located). From these definitions we may extend presence to telepresence by quoting Durlach[3] who explains it as a "subjective sensation of being present in a remote or artificial environment but not in the surrounding physical environment".

Yet, if we remember older technologies like the telephone, the television and even the fax, a possible perception of remote presence is also possible even when reading a book where we may be "transported" to a different place and/or to a different time. In fact, Gerrig[4] claims that telepresence may be considered as a sort of transportation where the human is taken to another location by some vehicle as a result of performing certain actions. As a consequence, the "traveller" goes some distance from his/her environment of origin making some aspects of his/her environment of origin inaccessible. If the trip has been a profitable one, the user will have learnt (distance education), controlled (remote monitoring) or enjoyed (entertainment) him/herself. Moreover, there are different sensations of place depending on the nature of the perception of presence which, in itself, is not generally continuous but a "moment-to-moment" feeling. We may have a presence related to a physical environment (the place where we are) which is a basic state of consciousness and may be considered as distal immediate. There is also a presence in a virtual environment, distal mediated, which is typically produced by the virtual reality technology, and another presence, in an imaginary environment, having a reduced attention to the distal stimuli, this is an internally generated mental imagery, not directly influenced by the media.

The importance of presence stems from its usefulness in a class of tasks which may only be performed if such presence is well perceived, maintained and user-independent; in fact many new possibilities emerge like tele-medicine, collaborative work, distance teaching, remote legal testimony, etc. which may only be performed on the basis of a "well perceived" state of presence.

DIFFERENT NOTIONS OF PRESENCE

There are a number of variations on the theme of presence, in fact it may be social, real, transported, immersed, enacted or medium-based.

Social presence implies a medium which provides warm, sensitive, intimate interactions between humans overcoming a number of constraints like time, permanence and distribution. Broadly speaking, social presence supports the transmission of symbolic and non-verbal human cues and, at the same time, reduce semantic ambiguities during the communication. There are a number of variables that play a role in social presence, they are cold/warm, impersonal/personal, insensitive/sensitive, etc. Moreover, we may speak of the richness of the information medium which is used for the creation of presence, such richness is based on the capacity of the medium for immediate feed-back (it is so disturbing

Figure 2. An artificial room with light effects (from Cornell University).

in a conversation to make an international phone call and hear the called person with a slight delay), the number of senses involved (visual, tactile, auditory, olfactory,...) and the personalization, including the language choice.

Which are the core communication elements of social presence? During close interaction in a local environment we may quote eye-contact, facial expression, voice quality, body language all contribute to closeness and must be allowed, at the right level of intimacy, in a recreated presence.

Real presence is one in which a medium can produce very accurate representations of objects, events and people, such representations must look, sound and feel as real as possible, the purpose here is in developing a perceptual realism that tries to match nature itself. See Figure 2 for a synthetic reproduction of an illuminated room.

Conversely, in a recent development area where agents are created for helping user in their work, we may speak of credible agents whenever they behave in a human-like way even if the representations involved are not realistic in any sense.

Transported presence tries to move the human from his original position in time and space to another place and has a well known ancestor which is fiction; nowadays we may see it in virtual tours of museums, in teleoperation of dangerous or heavy objects, etc. There are three variations of this form of presence: a) "you are there" which is the one we have just described, b) "it is here" whenever objects, scenarios, etc. seem to reach us as in IMAX films, wide screen projections, etc. and c) "we are together" where the communicants share a common space like in video conferences or in programs supporting a blackboard (shared window) where the different members may contribute in writing or sketching.

Immersed presence is a combination of perceptual and psychological immersion obtained by head-mount displays, headphones, gloves and even suits, the real world is invisible and only signals from the artificial scene may be received. The psychological components of this presence are involvement, absorption, engagement and engrossment.

Enacted presence relies on actors playing a script to produce the wanted effect: skilled personalities in TV enable para-social interaction using informal speech patterns, sincerity and simplicity which is furthermore enhanced by direct camera views or by camera movements which are not steady. In fact, media users (TV viewers) respond to social cues since the mediated nature of the interaction is ignored and the media personality is incorrectly perceived as a social actor. In this direction, avatars (incarnations of human form) like Mario, a sports mascot or Tamagotchi, a cyber pet, have been designed as interactors to humans in different, remote scenarios.

Medium-based presence is created for instance by programs, where users respond to medium cues so that a program is considered a social entity (a recognition program will be "correct" or "wrong" according to its performance). Some programs handle natural language, have some degree of undeterminism, interact in real time and fill traditionally social roles like a bank teller, a teacher, a schedule keeper. In electronic mail messages a politeness scheme is generally used, referred to as "netiquette", which stems from the media; gender stereotypes are also considered like evaluating a program instead of its programmer.

In conclusion, there are many different ways to support presence which is a property of a person and may vary in time and with the individual. Three factors play a role in the production of presence. Media form, content and users are all responsible, in some way, for the resulting effect of presence.

In fact, media form implies its richness and vividness, the number of sensory dimensions presented simultaneously, the sensory depth or resolution within each of the channels, the number and consistency of sensory outputs (audio, video, touch, smell,...) even if the visual and aural senses dominate with respect to the other remaining ones.

Within the visual component, the display features are very significant: image quality and size, proportion of visual field, motion and colour, 3D simulation on the plane and camera techniques like close-up, direct address, subjective take, shaky motion. On the other hand in aural presentations there is a mixed evidence about the obtained realism, in some cases, noise may play a positive role with respect to realism so that ambient sounds may help.

There are also other stimuli like olfactory (Sensorama, Plyester films) which are difficult to control and deliver. Body movement is important in specific film presentations like Earthquake and Rollercoaster (Sensoround system) to involve the spectators in a full tactile experience. The tactile sense is used in joysticks, finger hole (like in the Phantom haptic feedback device) where the illusion of a spongy or rocky material is provided by means of a sensation of physical resistance. The real problem with all the different senses used is one of integration so that the most important senses are used accordingly.

INTERACTIVITY

The essential feature that links all the different approaches to the creation of the feeling of presence is the possibility, for a user, to close a loop with the environment he is perceiving so as to feedback his commands to remote objects (teleoperation) or move around the created scenario (virtual navigation) or even change the point of view of an object on a screen (by moving the object on the screen in the opposite direction of the head movement of the observer when trying to see the left/right part of the object) or by zooming in a particular location when approaching it. In all these cases we need a fast feedback for providing interactivity between the user and the virtual reality in which he is projecting himself. Interactivity may also impinge on contents (not only on views) and the medium used may be described in terms of the number of inputs (for speech, for pressure, for gestures or other cues) and on the number and type of presentation features like temporal ordering, spatial organisation, intensity and frequency.

Each one of these features may have a given range (expressing the amount of possible change) used for a wide panning or deep zooming and, at the same time, may also have a given correspondence between the type of user input and medium type (familiar sensorimotor skills manipulating virtual objects directly may strongly contribute to our sense of presence). Finally, the speed with which the medium responds to the user's actions (possibly real time) is crucial in determining the feeling of presence and, for instance, in asking to users their preference between image resolution and motion responsiveness, they replied that this last feature was the most important one.

At the same time the medium should not be obtrusive to the objects that must be seen so as not to remind the user of a mediated experience; no edges, corners, keyboards, controls, lights, etc. should appear in the landscape. The presence of noise (both visual and acoustical) is discouraging in terms of the illusion of presence, and live scenes versus recorded ones are certainly better in terms of credibility.

USER VARIABLES

Since presence depends on the user it is important to analyse which are the user variables that must be considered. Firstly, willingness to suspend disbelief, so that presence is encouraged by decreasing awareness and this is a voluntary action, secondly since an a-priori experience knowledge may influence the feeling of presence, in this sense, experts are less prone to feel presence since they have already gone through the experience and they will easily detect system flaws. Another variable is related to learning skills which are very important in using the medium as when a joystick is used to navigate in a virtual

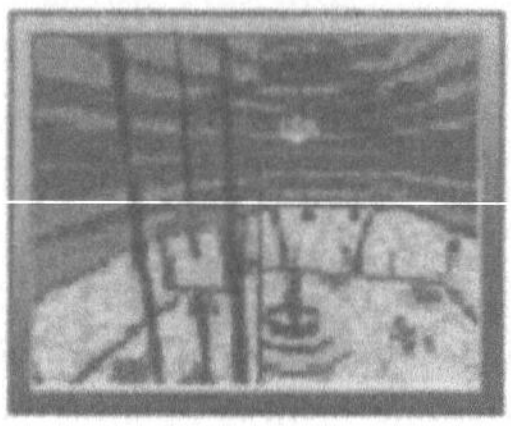

Figure 3. A virtual lift (GVU Centre, www.cc.gatech.edu/org, Georgia Institute of Technology).

reality scenario, not always an easy job. Finally, the personality type, which has a definite set of preferences (like infomercials, visual, auditory or kinaesthetic representations) and particular moods, ages, gender will also influence the presence effect.

MEDIA USERS

Much research has been carried out on the different classes of users and this for a number of reasons, because software products must address particular user communities, because to perform usability studies user models are required and also due to the need to teach computer science methodologies which will be applied to real software used by real people.

Age (children differ from adults in interests, skills and expectations) is important, particularly at the interface level where the medium is set, controlled and used. A number of physiological effects are very important like arousal, vection (illusion of self-motion) and phobias, flinching, ducking, flashbacks, graping arm rests, are some of the consequences of experiencing presence of different sorts. All these issues must be taken into account particularly when designing simulation systems to train pilots, surgeons, air traffic controllers.

A number of diverse effects of presence may be used for therapy like in the case of treating phobias where an artificial lift, created by means of virtual reality at the GVU Centre, see Figure 3. The view from the lift is perceived by patients and, by means of a desensitisation technique, after a number of sessions the patient starts to decrease his fear of high.

Another important effect is the one induced by infomercials where the message through a medium is provided in such a way as to recreate a real situation in normal life, this effect is one of persuasion and has a relevant economical value. Experiments designed to measure memory retention with respect to the representation of information have been performed and it was noted that when the information source was hidden, therefore providing a high presence effect, social judgement and memory retrieval capacity were notably increased. Some of these experiments merely compare the same information contents when presented on a small black and white TV screen with a high definition, colour, 52 inch screen.

USABILITY

There are many definitions of the term usability in the current literature, but historically, the first definition - even if a rough one - is attributed to Miller[5] - who intended "a property by means of which a computer system may be easily used". Only twenty years later it was recognised that four different factors are crucial in the definition and evaluation of usability. These are: effectiveness, learnability, flexibility and attitude. The first one deals with the matching between the task and the procedure to achieve it by means of the given system, the second one relates to the difficulty, for the user, in achieving his task using the system, the third one takes the degree of elasticity offered by the system to the user and the fourth one evaluates the user reaction (tiredness/frustration) after the goal has been reached.

Later on, a more comprehensive definition was provided, namely "as the extent to which a product can be used by specified users to achieve specified goals with effectiveness, efficiency and satisfaction in a given context of use". Such a definition has been brought to the forefront of software design because, for the first time, the user has been given an important role in the software production cycle. In fact, the new approach to the evaluation of the quality of software products (and even hardware ones) was seen from the user's point of view. This approach has coined new terms like user-centred design and even co-participatory design meaning that, for a given set of tasks, the user is the most expert person in the whole design team and should be involved in the inception of the product from the beginning. A very interesting book on usability and its different aspects together with the different possible ways to assess it is Nielsen;[6] software quality evaluation can be found in Bevan and Macleod.[7]

In Europe, the MUSiC standards (**M**etrics for **U**sability **S**tandards **i**n **C**omputing, Esprit Project 5429) were developed within the European Community Research projects and with contributions from the following member countries: Spain, Germany, Netherlands, United Kingdom, Italy and Ireland. These standards have shown that usability testing is quite a different activity from reviewing or demonstrating a prototype system to typical users which is considered a selling job. Evaluating products from a usability point of view implies that the user-program interaction is also included within the measurements.

In order to perform real usability measurements on products, with special reference to ISO 9241 (November 1995), it is important to recognise four different phases: inspection, observation, instrumentation and questionnaires. It is therefore mandatory to use an adequate methodology (coded in different recent standards) to integrate the data coming from all the four phases.

More specifically, if the usability is seen within the more general context of the quality of a product or service, we may consider the following definition of usability (ISO 9126): the set of functional and non functional properties, which are relevant to the productivity and satisfaction of the final user. There are a number of factors which may influence the usability feature, they are shown on Figure 4. As it can be seen, the quality of the interaction is a direct consequence of the coordinated reached levels in the effectivity,

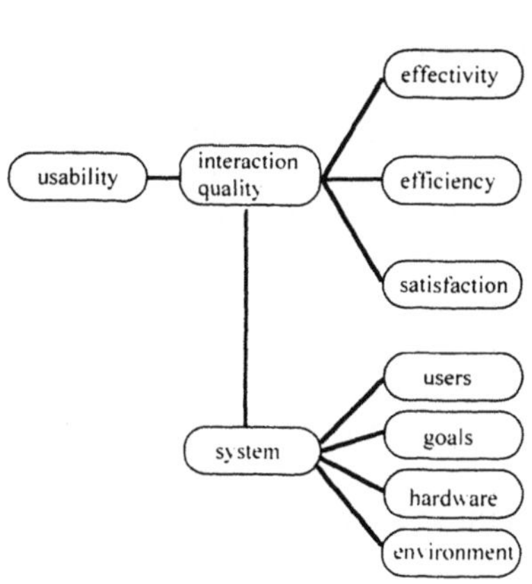

Figure 4. Usability and its components.

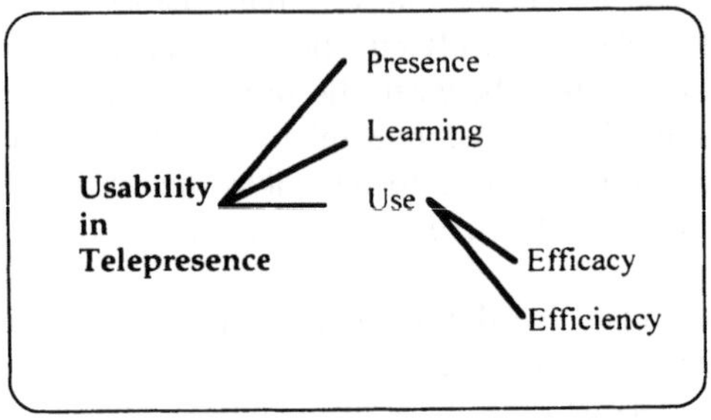

Figure 5. Different issues related to usability in telepresence.

efficiency and satisfaction (for the user) achieved by using the system.

Turning now to our specific field of interest, namely telepresence (in a number of applications including telemanipulation), we may consider its specific usability as made up of three different factors, see Figure 5. The need for a measurement of the effects of presence focusing on realism, possibility to change the observer's point of view and independence from the real physical position of the observer, is obvious. The sketch of Figure 5 underlines the dependency of telepresence from presence (there cannot be a good remote representation if the user does not feel stimulated by the senses so as "to be in another place" - presence). Furthermore, the difficulty/easiness of learning how to use the system and the results obtained with respect to the efforts employed (efficacy) as well as the good match between wanted goal and reached goal (efficacy) are the usability components in telepresence.

INTERFACE PROTOTYPE

We are at the beginning of a project which is aimed at the design of an interface for telemanipulators of mechanical robots so as to provide both virtual imagery and remotely sensed imagery (via television cameras) to the human operator, for training and working sessions as well as to enable comparison and integration of the information received to help achieve the task more efficiently and effectively.

Together with Infobyte, an Italian software house specialised in Virtual Reality applications, we have designed a first interface prototype which tries to address most of the issues mentioned in this chapter, particularly with respect to the operator interaction with the natural and artificial images which will be available during the telemanipulation process.

As a first example of the application of the guidelines, we show in Figure 6a, a few icons which satisfy the specification levels indicated by Crampton Smith,[8] which include concept, representation form, pictorial form and pixel form.

Our scroll bar represents a possible vertical shift of the darker rectangle, it has a smooth form enabling movement, an elementary pictorial shape which is easily translated into a pixel configuration requiring little detail.

The computer monitor icon, is too general (it meant "operation under computer control") and will probably be changed into another one standing for a more specific concept in the next version of the icon lexicon.

The slider has a very simple structure (a moving cursor on a black horizontal bar) with the extreme numerical values at the ends of the bar; there are no substantial differences between its pictorial representation and its digital one using pixels. In order to see the real

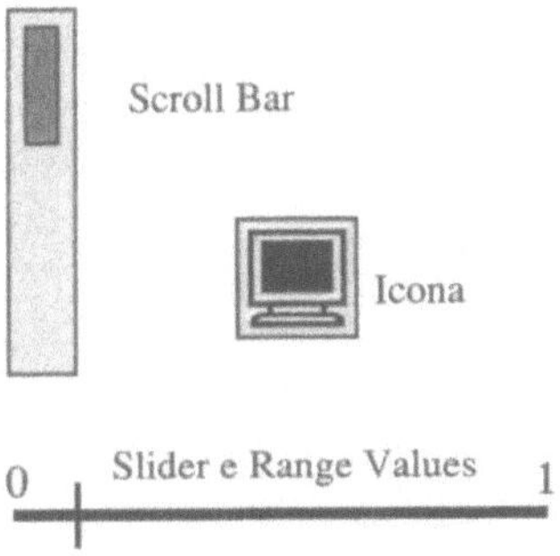

Figure 6a. Examples of icons.

Figure 6b. Range value.

value at which the cursor is positioned, a square window containing such value has been added as well as the parameter which is being modified by the slider (see Fig. 6b).

As may be seen in Figure 7, a large top picture on the left corner (containing the virtual reality image of the two robot arms and an object on a table) may be seen together with other five different views (always in artificial reality renderings) of the left, top, right, front and back directions with respect to the working area. A small cube (see bottom left corner of the first large image of Figure 7) is used to code, with one single letter, the viewing directions which are reminded to the user by colouring the frames of the corresponding views with the same cube face colour (e.g. the top is blue and the framing of the top view is also blue).

The right hand side bar of icons shows the active ones by a vivid colour (reddish) and the passive ones by a neutral, embossed profile of the same icon. In Figure 7 only the right brake is active meaning that the operator may only provide a breaking force (by means of a pedal) to the right arm. A warning window area is at the bottom of the screen frame to contain messages accounting the handling of exceptions and out-of-range values during telemanipulation.

In Figure 8 the same prototype interface is seen when a specific operation mode called "teach and repeat" (similar to a macro function made of elementary operations) is active. Only two small windows are present: the top and back view of the working area, the only active break is the right one and the operation mode is signalled by means of an icon representing an audio cassette (for a recorded sequence of actions). A number of possible

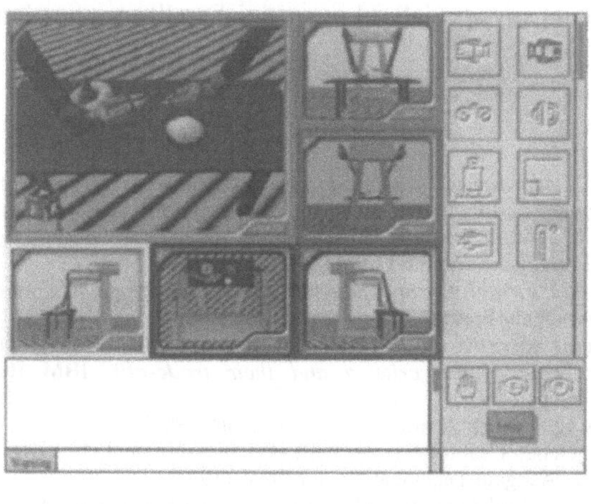

Figure 7. A first interface prototype (Infobyte™).

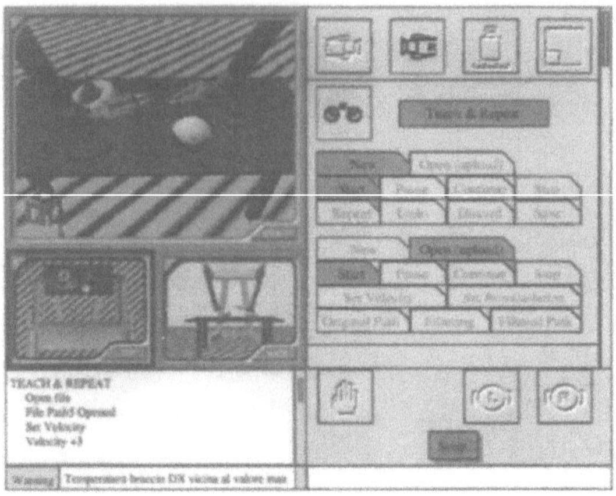

Figure 8. A view of the same prototype interface during a "teach and repeat" operation mode (Infobyte™).

actions are viewed by means of active cards (new, start, open, ...) and the set of operations which are performed, can be seen as a list on the left hand side (under the shown virtual images), the warning window contains a message related to the maximum allowable temperature of the right arm which is presently reaching its limit.

CONCLUSIONS

This chapter has reviewed the different ways of considering presence within information systems allowing the users to feel and act in a different place from the one they are presently standing. The possibilities which opens to human users are numerous, from telemedicine to teleoperation, from entertainment to interactive work with other cooperating members, suggesting that this technology has a great future provided some important human features are recognised, employed and trained to enhance task achievement and, more technically, to increase usability. This recent concept is covering all aspects of computerised systems particularly when they interact with users, so that both presence and telepresence fit into this general framework. Finally, a project for the design of a new prototype interface within a teleoperation environment, is very briefly introduced in terms of its graphical components and iconic metaphors. A few examples of screens are provided so as to give the flavour of the project itself which is carried out together with Infobyte, an Italian company specialised in artificial reality applications.

REFERENCES

1. T. Kim and F. Biocca, *Telepresence via television: Two Dimensions of Telepresence May Have Different Connections to Memory and Persuasion*, JCMC, http:/jcmc.huji.ac.il/vol3/issue2/kim.html (1997).
2. M. Lombard and T. Ditton, *At the Heart of it All: The Concept of Telepresence*, JCMC, http:/jcmc.huji.ac.il/vol3/issue2ombard.html (1997).
3. I. Durlach Nathaniel ed., *Virtual Reality: Scientific And Technological Challenges,* National Research Council, National Academy Press (1994).
4. R.J. Gerrig, *Experiencing narrative worlds*, New Haven, Connecticut, Yale University Press (1993).
5. R.B. Miller, *Human ease of use criteria and their trade-offs*, IBM Report, TR00.2185, IBM Corporation, Poughkeepsie (1971).
6. J. Nielsen, *Usability Engineering*, Academic Press, Boston (1993).
7. N. Bevan and M. Macleod, Usability assessment and measurement, in: *The management of software quality*, M. Kelly ed., Ashgate Technical/Gower Press (1993).
8. G. Crampton Smith, *Icon Guidelines*, Visual-L Discussion List, gillian@rca.ac.uk, October, 1998.

NON LINEAR MODELS FOR SPEECH MELODY

Roberto D'Autilia

Dipartimento di Fisica, Università di Roma "La Sapienza"
Piazzale Aldo Moro, 2
I-00185 Roma, Italy
e-mail: dautilia@romagtc.roma1.infn.it
http://romagtc.roma1.infn.it

INTRODUCTION

"Come hither, renowned Odysseus, great glory of the Achaeans. Stop your ship, so that you may hear our voice. No one has ever sailed by this place in a black ship until he has heard the honey-speaking voice that comes from our mouths; and he receives both delight and new knowledge before he goes on. For we know all the sufferings endured by the Greeks and Trojans in broad Troy at the will of the gods, and we know everything that happens upon the fruitful earth".[1]

By these words, pronounced by the Sirens in the Odyssey, Homer describes the great fascination the human voice exerted over the ancient Greek culture. Although our civilisation is quite far from that ancient one, speech is still the prevailing way to communicate "non semantic signals". However, in spite of its dominant role among acoustic signals, speech's physical nature is still far from being understood.

The purpose of this chapter is twofold: to describe an algorithm for the extraction of the prosodic patterns from speech and to show how it can be used for the analysis of ancient Greek prosodic systems and their connection with the development of languages. We want also to show why it is necessary, to this purpose, to renounce to the Fourier transform approach and to develop a complete and consistent theory of speech. Since the prosody is a non linear dynamic process, the Fourier transform, as well as other traditional speech analysis tools,[2] is conceptually inadequate to grasp the physical nature of speech melody. This becomes evident when we try to synthesise the speech by superimposing harmonics to obtain a signal that does not have any human feature.[3]

Although the computer time to run the Fourier transform is so small to make possible the completion of a great number of operations for unit of time, however it is really naïve to think that it would be possible to capture the non linear nature of the speech by means of a linear algorithm. In other words, in spite of its velocity, the Fourier transform is a backward theoretical step with respect to the physical modelling of the phenomena.

From an epistemological point of view, we observe that the Fourier approach is analogous to the epicyclic approach to the gravitation theory which considered the planets' orbits as the composition of circular motions. Ancient astronomers were forced to observe the planets' orbits for very long times in order to find all the elementary circular motion needed.[4]

In a similar way the modern engineering exploit the possibility to make a huge amount of arithmetic operations in a very short time to develop a speech synthesis and analysis by means of harmonics superposition. On the opposite side of this filtering mentality, the Newtonian approach developed to explain the gravitation suggests to find the equations for the motion of the celestial mechanical system.

Following the Newtonian point of view we want to find the equation of the motion for the speech. In this chapter we describe only the first step in this direction, which consists in identifying what are the fundamental variables of the phenomenum. Our main hypothesis is that speech, and in particular the intonation of the vowels, cannot be simulated just by filtering a fundamental buzzer. We make this point clear when we explain the physical model underlying the proposed analysis algorithms.

The non linear physical model upon which we construct the algorithm we propose is the behaviour of the self-oscillator. Speech is a dissipative process where the glottis and the buccal cavity dissipate the energy feeded by the air pushed by the diaphragm. The self-oscillating system given by the coupling of these two apparatus gives the speech articulation. For this reason it is not really possible to consider the frequencies of the speech, but instead it is necessary to deal with limit cycles, which reveal the main difference between the harmonic oscillators and the self-oscillating behaviour.

In light of this modelling objective the first step is to identify the variables of the phenomenum. To this end we observe that the articulation of speech melody is a process involving only the glottal part of the phonation apparatus. The sound of the vocal cords can be recognised when we listen to a speaking voice, and, in the frame of the characteristic glottal period, this motion is marked by the absolute maximum (minimum) of the amplitude. It is possible to define a function $p(t)$ which represents the time evolution of these peaks. This function is the object to study for the purpose of the recognition and synthesis of the prosodic patterns. From this function the new tool of the "prosodic notation" can be developed for the speech analysis. We use it for discussing possible connections of the ancient Greek prosody with coeval musical systems.

We are particularly interested in the musical and poetical systems of the ancient Greek civilisation and their assimilation in our musical systems. On the one side we want to understand if the prosodic notation can be used for studying the complex relationship between the speech and the singing in the ancient Greek musical-prosodic system, on the other to formalise the intermediate form between these two. This intermediate motion between the speech and the singing (the poetic declamation[5]) is one of the most interesting and least understood features of ancient literature and is connected with the understanding the "syllabic quantities".

The chapter is organised as follows: in the first section we explain the algorithm for the glottal pulses analysis and introduce the prosodic notation, showing qualitatively why a controlled self-oscillating system is a good model for the problem we are considering and why it is a general paradigm for speech. Then we discuss the relationship between the speech and the singing and discuss the possibility to interpretate the ancient Greek theoretical works on this subject by means of the prosodic notation introduced in the third section. At the end we suggest some future directions for this research.

THE SPEECH MELODY

We want to focus our attention on the complex task of modelling the melody of speech. There are two main reasons why we are interested in the study of the prosodic patterns: on the one hand we know that the speech melody holds the essential information we deliver when we speak, at least all the non-semantic information, on the other we suspect that the fundamental mechanism of the glottal pulse is a general non linear paradigm for the vowels articulation also. The second point will be developed in another paper,[6] here we want to focus our attention only on the analysis of speech melody. We do not want to study here the way the brain processes the acoustical signals, but only the dynamic process which produces the information perceived by the brain as the melodic articulation of the speech.

Speech melody is a non-semantic source of information, the emotional part of linguistic communication. This process is still far from being understood, and the lack of the prosodic features is one of the reasons for the failure of development of artificial speech. Indeed prosody is not only an important source of information but also a fingerprint for the way of talking for a group of people. Only a fraction of the prosodic information is marked by means of the punctuation marks, but the number of prosodic patterns is really huge. Even the simplest prosodic formula, such as the end of an interrogative sentence, corresponds to many different melodies. If we want to make a systematic study of these formulas, we need a non ambiguous definition for prosodic patterns, a signal which holds only the prosodic features.

Let us call $s(t)$ the signal given by the speech in the time interval $[T_0, T_1]$, the time evolution of the air pressure

$$[T_0, T_1] \ni t \rightarrow s(t) \in [-1,1] \qquad (1)$$

normalised to fit in the [-1,1] interval.

The signals we analyse were recorded by means of a dynamic unidirectional microphone with a 50-15.000 *Hz* frequency response and discretised by means of a 20.000 *Hz* Analog Digital Converter.

In Fig.1 the evolution of the function $s(t)$ with a duration of 0.15 seconds for the "A" vowel pronounced by an adult male is shown.

Since the brain cannot elaborate so many data in such a small amount of time, signals like this "A" hold a very large quantity of information, so that it is realistic to think that only a small part of it is relevant for prosodic communication. In other words we need to extract from $s(t)$ a new signal

$$[T_0, T_1] \ni t \rightarrow p(t) \in [-1,1] \qquad (2)$$

holding all the prosodic features of $s(t)$.

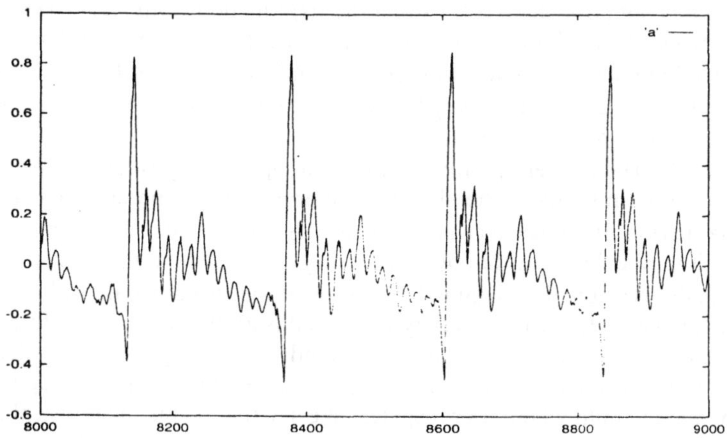

Figure 1. The $s(t)$ signal corresponding to the "A" vowel.

Let us define the function $m(t)$ as the time evolution for the peaks of $s(t)$:

$$[T_0,T_1]\ni t\rightarrow m(t)=\begin{cases}s(t) & \text{if } s(t) \text{ is a local maximum}\\ 0 & \text{otherwise}\end{cases} \tag{3}$$

then given a number α, we select a peak at time t if $m(t)$ is greater or equal to the other peaks in the $[t-\alpha,t+\alpha]$ interval, and if the time elapsed since the last selected peack is greater than α

$$[T_0,T_1]\ni t\rightarrow p(t)=\begin{cases}m(t) & \text{if } m(t)\geq m(z),\ z\in[t-\alpha,t+\alpha] \text{ and } \Delta>\alpha\\ 0 & \text{otherwise}\end{cases} \tag{4}$$

where Δ is the time elapsed since the last selected peak. The local maxima for the $s(t)$ in the $[t-\alpha,t+\alpha]$ interval can be detected with a delay not greater than α, so that the algorithm can be implemented in real time. A simple computer algorithm for extracting $p(t)$ from the $s(t)$ is the following. Let us discretise the time, T_0, T_0+1, ..., T_1-1, T_1 and define the set

$$Z=\{z\colon s(z)\geq s(z+k)\quad \forall\, k\in[1,\,1+\alpha]\} \tag{5}$$

The $p(t)$ can be obtained by means of the following recursive algorithm

$$\tau_0=\min\{z\in Z: z\geq T_0\} \tag{6}$$

$$\tau_1=\min\{z\in Z: z>\tau_0+\alpha\} \tag{7}$$

and in general

$$\tau_i=\min\{z\in Z: z>\tau_{i-1}+\alpha\} \tag{8}$$

Then $p(t)$ is given by

$$p(t)=\begin{cases}s(t) & \text{if } t\in T=\{\tau_0,\tau_1,\ldots\}\\ 0 & \text{otherwise}\end{cases} \tag{9}$$

The set of the non-zero values for $p(t)$ is very small if compared with the discretised $s(t)$, and of course it is not possible to reconstruct the $s(t)$ function from $p(t)$. The free parameter α, a characteristic of the glottal period, is the average distance between two neighbouring peaks. For an adult male α is about 0.08 sec and can be easily identified at the beginning of the analysis. In Fig. 2 the function $p(t)$ with $\alpha=0.08$ is plotted together with the $s(t)$ corresponding to the "A" vowel of Fig.1.

We selected $p(t)$ considering only the maxima, but the algorithm can be improved by considering also the minima of the signal. Furthermore the α parameter is related to the proper glottal period of the speaker and can vary during the speech. In another paper[6] we show how to improve the algorithm by considering a time depending of $\alpha(t)$.

Although the $p(t)$ function obtained by means of the recursive algorithm is a sequence of spikes, it is nevertheless easy to verify that it holds all the prosodic features of $s(t)$. We found that the signal we perceive as the melody of the speech is a discrete signal corresponding to the main peaks, a sort of percussive signal whose shape is generally smoothed by the medium where it propagates. We say that $p(t)$ is the physical definition of the prosody. We also observe that if we use the $p(t)$ function for the additive synthesis we

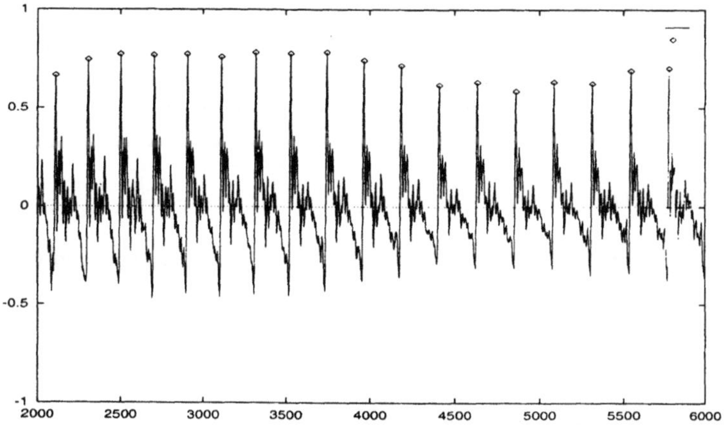

Figure 2. The $s(t)$ and $p(t)$ functions for the "A" vowel.

have better results than those obtained just filtering a regular signal.

The values of $p(t)$ have been selected on the basis of the signal amplitudes. We want briefly to discuss why the selection of the peaks on the basis of the amplitude is a fundamental feature of the dissipative characteristics of the self-oscillating systems. Without entering in the details, we only give a qualitative account of this. We recall that a physical model is an algorithm for making predictions about a given phenomenum. These predictions, output of analytical or numerical computations, are numbers which have to be compared with the experimental data. In the modelling approach the analysis and the synthesis are two aspects of the same knowledge process. When we analyse the signal we identify the parameters of a given model, where, during the synthesis, the model runs with the identified parameters to give data that are coherent with the experiments.

The physical system underlying the proposed analysis is the self-oscillator:[7] a dissipative system which, feeded by an external source of energy, exhibits an oscillating motion. Physical examples of self-oscillators are the electric bell, the relaxation oscillations, the steam engine and the bowed strings.[7] Speech is a dissipative system because of the dissipative nature of the glottis, and the glottal pulses is also an example of a controlled self-oscillating system. Indeed if we push the air in the glottis by means of the diaphragm, the vocal cords are initially closed. The action of pushing increases the pressure behind the glottis up to a threshold controlled by the muscular tension. When the pressure is higher than the threshold the vocal cords suddenly open and the air flows out. With the open vocal cords the diaphragm pressure decreases, the glottis close again by the muscular tension, and a new cycle starts.

The peaks corresponding to the maximum (the minimum if we invert the microphone polarity) mark, in this qualitative picture, the instants when the glottis is open. The strength of the glottis (the threshold) is controlled by the muscles and increases or decreases the time between a peak and the following one. The slowing or the speeding of the peaks process gives the prosodic articulation. We notice that the concept of frequency cannot be used because the $p(t)$ motion is not a periodic motion. It is the non periodicity of $p(t)$ the main feature of the speech which gives the possibility of using this system for communicating information.

Many of the features of the self-oscillators, such as the functional relationship between the glottal period and the strength of the input force, can be detected analysing $p(t)$. If we scream we observe an increase of frequency which is a feature of the limit cycles of the self-oscillating systems which cannot be modelled in terms of harmonic oscillators. It is interesting to notice that this connection between amplitude and frequency was well known to Greek science,[8] already at the time of Pythagoras. We want to observe that the mechanism of the glottal pulse could be extended to the vowel articulation,[6] although we do not discuss this point here. In the next section we introduce the prosodic notation, and show why it can be a powerful analysis tool for study of the prosody.

THE PROSODIC NOTATION

Let us introduce the time evolution of the temporal distance between two neighbour peaks of the $p(t)$ function

$$[T_0, T_1] \supset T = \{\tau_0, \tau_1, \ldots, \tau_N\} \ni \tau_i \to d(\tau_i) = \tau_i - \tau_{i-1} \equiv d(t) \qquad \text{for } i \geq 1 \tag{10}$$

The $d(t)$ is a function of the τ's, but since the process of extracting the τ's depends on the $t \in [T_0, T_1]$ we prefer to use the notation

$$[T_0, T_1] \ni t \to d(t) \in \Re^+ \tag{11}$$

to draw attention to its temporal dependence. In Fig.3 the $d(t)$ function for the vowel "A" is plotted, where the units for $d(t)$ are 10^{-10} seconds. The points are joined with lines only for the purpose to stress the prosodic evolution, but the $d(t)$ function is a discrete function.

It is clear the pointwise variability of this function which gives a graphical representation of the prosody. For this reason we call "prosodic notation" the function $d(t)$. The $p(t)$ process is a controlled self-oscillating process, where the $d(t)$ function is a plot of the time evolution of the control. The values of the $d(t)$ points are about 10^{-2} seconds which is the characteristic of the brain and the muscular control. It is important to observe that the function $d(t)$ changes its values almost in every point with respect to the previous non zero value.

Because of this pointwise variability, the Fast Fourier Transform (FFT) cannot be used to detect the $p(t)$: we would need a temporal window larger than the distance between two neighbour peaks to detect a frequency. For this reason the F_0, the fundamental of the FFT, can give only a very rough sketch of $d(t)$ and does not detect the microvariability of the speech intonation which is also a discrete signal and not a continuous one. It has to be noticed also that $d(t)$ for the singing would give orizontal segments.

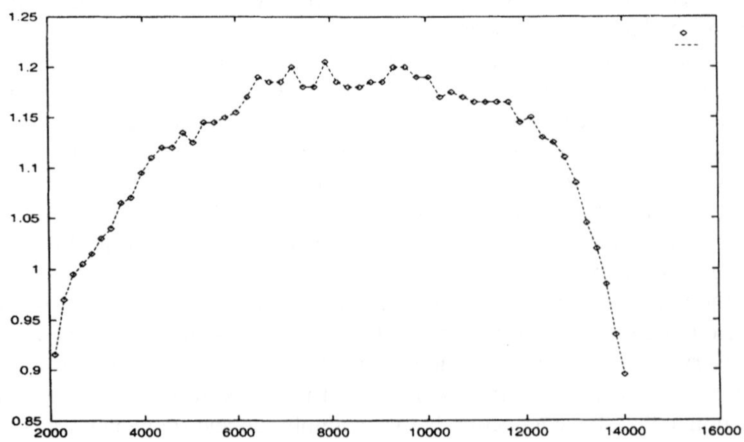

Figure 3. The temporal distance of the glottal pulses for the "A" vowel.

Of course it is also possible to plot the function $1/d(t)$ which is a kind of frequency evolution, but the concept of frequency is misleading because we are not dealing with periodic processes. We can apply, for example, the proposed analysis to the output of a damped oscillator to produce a $d(t)$ function which is a set of points with the same values, where the Fourier Transform of $s(t)$ does not give any frequency.

For the purposes of historical analysis we prefer to use in the next section the $-d(t)$ function for its physical and physiological meaning. The $d(t)$ signal can be used to make a vowel segmentation of the speech, since it shows a sharp distinction between the very regular behaviour of the vowels and the disordered behaviour of the consonants. In a next paper[6] we show how to classify the regularities of $d(t)$ to recognise the vowels.

The function $p(t)$ is the prosody and the $d(t)$ is the plot of its control. The two functions capture the non-linear features of the peaks control, coherently with the strong non-linear characterisation of the speech which is not detected by the epicyclic linear analysis. In the next section we see how to use the prosodic notation to make some considerations about the neumatic notation of the Gregorian chant, and suggest a possible relationship with the prosodic notation.

ANCIENT MUSIC AND ANCIENT SPEECH

The algorithms leading to the $p(t)$ and $d(t)$ functions are founded on the general principle that the controlled self-oscillator is the universal tool for the acoustical communication. Many musical instruments, like string or wind instruments, exploit the self-oscillations to produce sounds. The behaviour of these instruments is the same of the glottis. To give an example, when the bow starts moving on the violin, it drifts the string until the tension is bigger than a threshold over which the string makes a dumped oscillation. Then the velocity of the string decreases (because of the dynamic friction) and when it is small enough it stops again making a new cycle starting. We can suggest that the self-oscillation is the common way of working for the speech and the bowed instruments, for the prosody and the music.

If we read the ancient Greek musical writers, we find that in the ancient Greek civilisation the sharp difference between the continuous motion and the intervallic motion for the speech was clear,[9] but the intermediate motion between speech and singing was used to read the poetical compositions.[5] The Greek musical writers who wrote about the intervallic and the continuous use of the voice, stressed the great emotive power of the mixed speech represented by the quantitative metrics.[10] It is also known that already five centuries before Homer in Greece there were minstrels chanting the poems accompaning themselves on the *phorminx*, and that from the middle of the sixth century the same texts (that represented the educational basis of that civilisation) were recited by the *rhapsodoi*. So we can argue that in ancient Greece a strong prosodic and musical awareness was quite common. We suppose that this sensitivity was the basis of a sort of prosodic notation which is at the origin of the neumatic and ekphonetic musical notations.[11,12] The music had been monodic up the middle age, and we can suppose that the syntax of the melody[13] is in some sense related to the prosody of the rethoric.

In other words we think that the neumatic notations are derived from a notation which is essentially the $-d(t)$ function for the speech, and that the *corpus* of ancient melodies originated from some kind of prosodic formulas. If we use the prosodic notation $d(t)$ to investigate the melody of the speech, we find in fact a natural neumatic notation for the speech.

Let us consider for example the beginning of the *Commedia* by Dante Alighieri:

Nel mezzo del cammin di nostra vita
mi ritrovai per una selva oscura
che la diritta via era smarrita

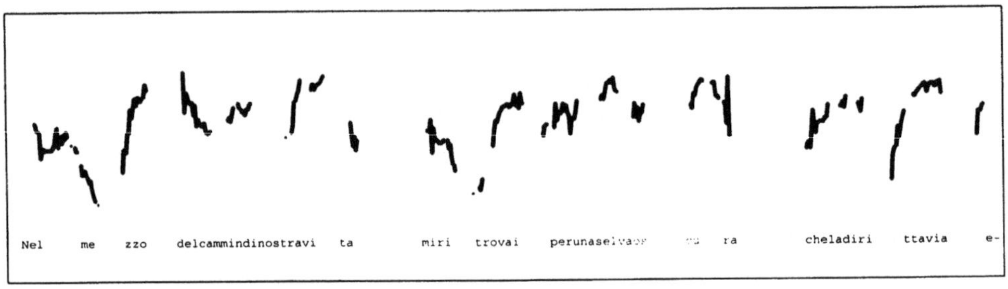

Figure 4. The $d(t)$ function for the beginning of the Dante's *Commedia*.

and apply the prosodic algorithm to extract the prosodic notation. We find that the prosodic notation marks the text to give the natural movement of the voice in a way which is similar to the medieval neums. For the purpose of highlighting the neumatic analogy with the prosodic notation, we do not show in Fig.4 any unit for the prosodic notation. We plotted only the $d(t)$ "segments" with a regular behaviour, eliminating all the $p(t)$ values under a given threshold.

If we make a comparison between the neumatic notation for the Gregorian chant (see for example the Einsiedeln cod. 121) and the $d(t)$ function, we find some analogies. We do not say that we found an algorithm for understanding the neums, but only that there is a natural neumatic notation for the speech, given by $d(t)$, and that it could be related to the origin of the neumatic notation. By connecting with a line all the points of $d(t)$ that are closer than a 2α interval we obtain notation for the interrogative cadenza "di nostra vita" shown in Fig.5 where the values of $-d(t)$ are plotted to stress the analogy with the neums.

From these analogies we propose the following possible interpretation for the origin of the musical system. The self-oscillating instruments could have been initially developed to simulate the prosodic patterns of speech, and later, when these patterns were well known and widely used, these instruments were connected to the resonators to stabilise the frequencies thus marking the difference between the prosody and the music. This linearisation process corresponds to the approximation of the $d(t)$ function by means of horizontal lines.

Many of the features of the non-linearity of self-oscillators which are exploited by the musical instruments (like for example the dependence of the frequency on the amplitude) conditioned also the aesthetic of music: just like the human voice, traditional musical instruments are louder in the high register than in the low, so that it is natural that in the *Zauberflöte* by Wolfgang Amadeus Mozart the *Königin der Nacht* and *Sarastro* sing on the opposite registers and that the *crescendo* in music is generally used for the ascending motion.

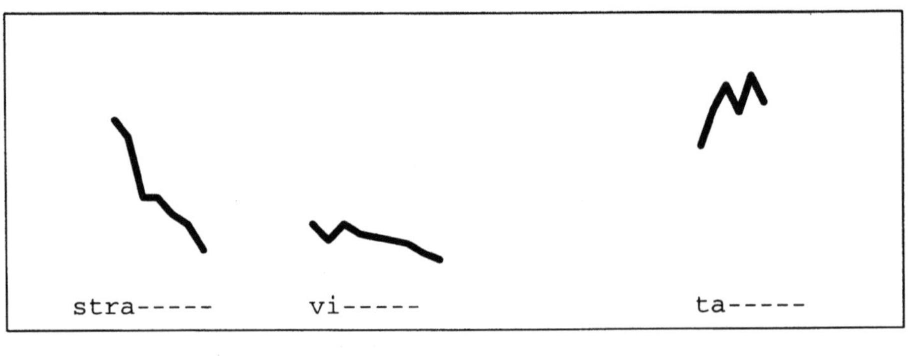

Figure 5. The $d(t)$ function for the "di nostra vita" from Dante's *Commedia*.

The hypothesis of the prosodic origin for music has been considered by Jean Jacques Rousseau,[14] and many musicians (see for example Janacek) tried to make a musical transposition of prosody. However has to be noticed that even the musicians who did not consider the prosody as a source for their music were submitted to the strong influence of the speech melody (see for example the Jews choir in *Salome* by Richard Strauss, where the Yiddish prosody is easily recognisable).

We suppose that the prosodic system is the origin of the music, that the modern musical systems are the evolution of the ancient prosodic system, and that even in modern times the prosody and the melody influence each other. There is not evidence that the ancient musical systems originated from the theory of the vibrating string, but instead there is evidence of a system of prosodic marks similar to the Gregorian neums or to the ekphonetic notation for the bizantine music.

In another paper[15] we analyse the possibility that music is a "ritualisation" of the prosodic patterns achieved by means of a linearisation process for the prosodical undulations analogous to the quadrate notation transcription of the neums. We suppose that at the origin of the Christian chant the declamation of the holy texts was a prosodic declamation, and that it was marked by means of a prosodic notation which became later the neumatic notation. After five centuries this notation shows a sort of linearisation. We think that this melodic transformation of the prosody happened several times in history, probably everytime some prosodic formulas became widespread they have been transformed in melodies.

We observe also that the neumatic notation is a continuous notation, more similar to the $d(t)$ of the speech than to the piecewise constant corresponding function for the singing, so that probably the musical transcriptions of the neums cannot be considered a translation, but only an arbitrary discretisation, derived by the melodic interpretation of the corresponding text. We suppose that this linearisation could have happened many times in the history of music, since the time of the application of the resonator to the self-oscillator, and slowly contributed to form our musical system.

An interesting question generated by our hypothesis is if the musical syntax also derived from a prosodic syntax, and if it is possible to find grammar and a syntax of the prosody by means of the $p(t)$ and $d(t)$ functions. Although our objectives do not include the study of the brain process related to the prosody, it could be interesting to analyse the possibility that the brain produces temporal connections among different prosodic patterns to develop a syntax of the prosody to be compared to the musical syntax. Another reason why it would be relevant to study the syntax of the prosodic notation, is that the speech melody conditions the evolution of the speech. Words in fact tend to be phonetically attracted to similar words with phonetic analogies which are distant from the semantic point of view. When analysed in the animal languages, prosodic syntax appears to be generated by a sort of conditional reflex among the prosodic patterns.[15]

A musical consequence of the prosodic origin for music is that the Greek musical modes could have been generated by the prosody of the different dialects. It is easy to verify that from the prosodic point of view the interrogative sentences or the cadences in the speech are different for various Italian dialects. It is possible that the prosodic studies of the ancient modes give indications about the difference among the ancient Greek dialects.

CONCLUSIONS AND ACKNOWLEDGMENTS

The main purpose of this chapter was to introduce an algorithm to extract the prosody from the speech, and to show why the detection of a microvariability for the peaks temporal distance can be an effective tool for esploring the speech melody as well as for the understanding of the melodic transformation of the prosody. We want to stress that this transformation is not given by the mechanism used for example by Janacek to compose his music, because it cannot be a simple melodic transposition of the prosody. Instead it is the historical assimilation of some prosodic patterns that generate "popular melodies" which form the *musical corpus*, which is the result of a very slow "percolation process" of the prosodic structures into the music. Probably the linearisation mechanism it is not the same in every time, the Greek transformation is different from the medioeval one.

The next step of this research is the study of the self-oscillating behaviour which produce the prosodic patterns,[6] where we consider also the modularity of the self-oscillators for the purpose of vowel recognition. We said that the prosodic patterns are fingerprints for the linguistic group, in this framework if we introduce a notion of distance between two prosodic patterns, the prosodic notation can be used to reconstruct the evolution of language for human beings which can be then compared with the corresponding genetic patterns.

The author would like to thank Carla Scaletti and Kurt Hebel of the Symbolic Sound Corporation who developed the hardware[17] and software[17] we used for implementing the $p(t)$ and $d(t)$ finction, Francesco Guerra for the indispensably fruitful discussions and suggestions, Nicola Cufaro Petroni for the observation about the prosodic notation, Gian Marco Todesco and David Calef for the revisal of the manuscript.

REFERENCES

1. Homer, *Odyssey* XII, 181-200, Andrew Barker, *Greek Musical Writings: I*, Cambridge University Press, Cambridge, Great Britain (1989).
2. I. Daubechies ed., *Proceedings of Symposia in Applied Mathematics* (Volume 47) - Different Perspectives on Wavelets, American Mathematical Society, Short Course, January 11-12, (1993).
3. Hermann von Helmholtz, *On the sensation of tone*, London: Longmans, (1885).
4. G. Van Brummelen, Lunar and planetary interpolation tables in Ptolemy's Almagest, *J. Hist. Astronom.*, 25 (4) 297-311 (1994).
5. Aristides Quintilianus, *De Musica*, R.P. Winnington-Ingram ed., Leipzig, (1963).
6. R. D'Autilia and F. Guerra, *Self-oscillating behavior of the speech signal*, in preparation.
7. A. A. Andronov, *Theory of Oscillators*, Pergamon Press, Oxford (1966).
8. A. Barker, *Greek Musical Writings: II*, p.41, Cambridge University Press, Cambridge, Great Britain (1989).
9. Aristoxenus, *Elementa Harmonica*, R. da Rios ed., Rome, (1954).
10. Nicomacus, *Enchiridion*, in C. von Jan, *Musici scriptores graeci*, Leipzig, (1895).
11. Carsten Höeg, *La notation Ekphonétique*, Monumenta Musicae Byzantine, Subsidia, vol.I, fasc.2, (1935).
12. Jean-Baptiste Thibaut, *Origine byzantine de la notation neumatique de l'Eglise latine*, (1903).
13. Fux, *Salita al Parnasso, osia Guida alla Regolare Composizione della Musica*, Forni Editore, Bologna (1972).
14. J.J. Russeau, J. G. Herder, *On the Origin of the language*, Chicago University Press, Chicago (1986).
15. R. D'Autilia, *Prosody and neums*, in preparation.
16. K. Hebel, C. Scaletti, *The Software Architecture of the Kyma System*, Proceedings of the 1993 Computer Music Conference, San Francisco: International Computer Music Association (1993).
17. K. Hebel, *A Framework for Developing Signal Processing and Synthesis Algorithms for the Motorola 56001*, Proceedings of the 1991 Computer Music Conference, San Francisco: International Computer Music Association (1991).

PANEL SUMMARY: "HOW CAN I KNOW WHAT I THINK UNTIL I SEE WHAT I SAY?"

Walter Gerbino

Department of Psychology
University of Trieste
I-34123 Trieste, Italy

HOT POINTS

- How do we get the feedback for our thoughts?
- Do people think better when communicating with others?
- Teaching: the best way of learning!
- Do people perceive the meaning of their words?
- Are personal and interpersonal symbols different?
- Do humans shift from personal to interpersonal symbols in everyday life?

Let me thank Virginio Cantoni for choosing this title and for inviting me to discuss such a stimulating topic. First of all I will try to clarify my way of understanding this question, that may be open to different interpretations.

It seems to me that the title poses a wonderful problem and formulates a possible suggestion for its solution. The problem regards the means for representing to ourselves the content of our own thoughts. The suggested solution rests on two ideas: (i) direct perception of the meaning of communicative acts is taken for granted; (ii) knowledge of our own thoughts is indirect, being mediated by perception of internal simulations of external communicative acts.

Let us accept the assumption that perception of the meaning of external communicative acts is a primary and direct form of cognition. One might disagree and claim that the meaning of a multisensory pattern such as the roaring and optically expanding mouth of a tiger requires the activation of a not-to-be-taken-for-granted cognitive machinery. However, for the sake of our discussion let us stipulate that humans and other intelligent systems can *see* the meaning of complex patterns of sensory information; i.e., they can utilise a direct form of apprehension of external events that, as a consequence, acquire the status of signs in social communication. In this context social communication encompasses any transmission of meaning between agents; by definition, it is involved in realising that the tiger is perceiving you as a succulent meal, in listening to sports news, as well as in talking to your best friend. The assumption that meanings mediated by language are picked-up directly is consistent with the phenomenology of communication. When you say "Yes" or "No" you expect to be immediately understood, and you behave accordingly. This holds for verbal and non-verbal codes; in both cases one can hypothesise the existence of specialised linguistic modules translating sensory patterns corresponding to signs into meanings.

Thinking poses a somewhat different problem. For "those who can read without moving their lips" (to quote a famous British ad) thinking is a private activity devoid of external perceivable correlates. If one just thinks "2+2= 4", "Walter is completely crazy", or "cognitive science does not exist", nothing happens unless such a person is connected to a fancy machine capable of registering brain activity with unbelievable accuracy and precision. Therefore, how can the thinker perceive the meaning of his/her thoughts? According to the idea suggested in the title, the problem does not apply to verbalisations. Thinking aloud does not constitute a problem because speech provides an obvious object for perception; whereas thinking alone would constitute a problem.

This line of reasoning drives us towards the following solution. I know what I think because I perceive my mental products. Note that we can distinguish between thinking as a form of implicit cognition and (partial or complete) awareness of thoughts. As suggested in the hot topic for our panel discussion - *How do we get the feedback for our thoughts?* - to direct our cognitive activity we need evidence at least about the outcome of processes that go on in our mind. An interesting possibility arises if the feedback consists in the availability of personal symbols grounded in perception.[1,2] Two theoretical aspects seems important to me. First, the idea that symbols used for mental representation and manipulation of meanings are personal; i.e., they may differ in different individuals though they are general enough to allow the production of communicable outcomes. Second, they are sufficiently general because humans and several other animals share a common world that constrained the development of perceptual systems in similar ways. The interplay between individuality and generality of mental symbol systems is at the heart of psychology as a natural science. To understand individual minds we should not overestimate the sharing of common ideas. People think different, in the sense that they use different tricks to keep in mind meanings and manipulate them. Nevertheless, people communicate because they can also use social codes. Probably, we do not know enough about the shifts between personal and interpersonal codes.

The second hot point for our panel discussion regards whether *people think better when communicating with others*. The issue is controversial. Some people claim that, to reach satisfactory conclusions, they need to stop, isolate from others, and think. Others claim to experience difficulties in thinking by themselves, and appear extremely productive in the context of social communication. As a contribution to our discussion, I would suggest that communication constrains thinking in ways that can be negative or positive, depending on conditions. Generating a verbal discourse or a pictorial illustration of a certain line of reasoning is a sequential procedure in which a single difficult step can slow down the full process. On the other hand, the communication framework can help by itself. For instance, to generate a good causal explanation we must order concepts and reproduce a chain that can be isomorphic to the sequence of events under consideration.

This is connected to our third point, summarised by the provoking claim *Teaching: the best way of learning!* It is a widely held belief that arguments are fully mastered only after they have been explained to other persons. The psychological truth behind this belief probably relates to the need of checking implicit assumptions hidden in personal modes of thinking. Thinking for ourselves is fast and phenomenally clear. But, as most of us know quite well, our own ideas often become less clear when we must articulate and make them available to others. Therefore, it is important to keep in mind that knowledge of our thoughts before their external expression might differ from knowledge reached, for instance, after writing. Psychologists cannot be impressed by the conflict between thoughts, on one hand, and verbal or behavioural expressions of those thoughts, on the other. The problem is not the existence of contradictions, so common in human minds. The interesting problem is the possible heterogeneity between knowledge based on thought-by-itself and knowledge based on public symbols.

Three related questions can be proposed as hot topics. *Do people perceive the meaning of their words? Are personal and interpersonal symbols different? Do humans shift from personal to interpersonal symbols in everyday life?*

The process of translating ideas into words appears to me as a fascinating business. You put the right question to a person and she can produce a long answer that seems to come out of nowhere with incredible easiness. We hypothesise that the question activates a complex distribution of connected meanings that the language production module articulates following an ordered path. Competent speakers generate fully adequate (sometimes optimal) verbal descriptions in a snapshot. This appears fascinating because in many instances it is clear that speakers have limited access to the range of meanings supported by a given verbal expression. Therefore, my answer to the question *Do people perceive the meaning of their words?* would be "Yes and no, it depends". The meaning of current actions and linguistic expressions is only partially represented in the mind of a communicating agent.

Our panel discussion supports the idea that an intelligent system capable of thinking (representing to itself) and communicating (representing to others) should integrate different symbolic modes. One possible architecture involves the coexistence of two devices: a simulator utilising private objects and a discourse/action generator that partially translates the simulated pattern into communicable symbolic chains.

REFERENCES

1. P.A. Kolers and W.E. Smythe, Images, Symbols, and Skills. *Canadian Journal of Psychology*, *33*, pp. 158-184 (1979).
2. L.W. Barsalou, Perceptual symbol systems. *Behavioural and Brain Sciences*, in press.

INDEX

MIX
Papier aus verantwortungsvollen Quellen
Paper from responsible sources
FSC® C105338

If you have any concerns about our products,
you can contact us on
ProductSafety@springernature.com

In case Publisher is established outside the EU,
the EU authorized representative is:
Springer Nature Customer Service Center GmbH
Europaplatz 3, 69115 Heidelberg, Germany

Printed by Libri Plureos GmbH
in Hamburg, Germany